Energy Centered Maintenance — A Green Maintenance System

Energy Centered Maintenance — A Green Maintenance System

Marvin Howell

Fadi S. Alshakhshir

Routledge
Taylor & Francis Group
LONDON AND NEW YORK

Published 2020 by River Publishers
River Publishers
Alsbjergvej 10, 9260 Gistrup, Denmark
www.riverpublishers.com

Distributed exclusively by Routledge
4 Park Square, Milton Park, Abingdon, Oxon OX14 4RN
605 Third Avenue, New York, NY 10017, USA

First issued in paperback 2023

Library of Congress Cataloging-in-Publication Data

Names: Howell, Marvin T., 1936- author. | Fadi S. Alshakhshir, 1982- author.
Title: Energy centered maintenance : a green maintenance system / Marvin
  Howell, Fadi S. Alshakhshir.
Description: Lilburn, GA : The Fairmont Press, Inc., [2017] | Includes
  bibliographical references and index.
Identifiers: LCCN 2016053056| ISBN 0881737798 (The Fairmont Press, Inc. :
  alk. paper) | ISBN 9781138735378 (Taylor & Francis distribution : alk.
  paper) | ISBN 9788770222662 (electronic)
Subjects: LCSH: Buildings--Maintenance. | Green products. | Industrial
  buildings--Energy conservation.
Classification: LCC TH3351 .H69 2017 | DDC 696--dc23 LC record available at
  https://lccn.loc.gov/2016053056

*Energy centered maintenance : a green maintenance system /Marvin Howell, Fadi S.
  Alshakhshir*
First published by Fairmont Press in 2017.

Routledge is an imprint of the Taylor & Francis Group, an informa business

Publisher's Note
The publisher has gone to great lengths to ensure the quality of this reprint but
points out that some imperfections in the original copies may be apparent.

10: 0881737798 (The Fairmont Press, Inc.)
13: 9781138735378 (print)
13: 9788770222662 (online)
13: 9781003151371 (ebook master)

While every effort is made to provide dependable information, the publisher, authors, and
editors cannot be held responsible for any errors or omissions.

The views expressed herein do not necessarily reflect those of the publisher.

ISBN 13: 978-87-7022-944-9 (pbk)
ISBN 13: 978-1-138-73537-8 (hbk)
ISBN 13: 978-1-003-15137-1 (ebk)

Dedication

From Fadi S. Alshakhshir

This book is dedicated to the memory of my father, to my mother, my family, my colleagues and my friends for giving me the strength to complete this work.

My appreciation and sincere gratitude are also extended to my influencer Mr. Abdulla Haji Mohamed A. Al Wahedi for his guidance and support throughout this time, and especially for his trust in my capabilities.

My thoughtful appreciation is also extended to my motivator Mr. Bashar Abu Fashah who believed in me and continuously supported me throughout my career life.

From Marvin Howell

This book is dedicated to Valerie Oviatt, Director, Seminars and Internet Training, AEE Energy Training who selected me as an AEE energy instructor and approved Energy Centered Maintenance (ECM) to be an on-line energy seminar.

In addition, I would like to thank my wife, Louise Howell, for her continued support.

Contents

Chapter 15

Preface

Energy Centered Maintenance (ECM) was originated in 2012 by Marvin Howell who holds a BS in mechanical engineering from Mississippi State University, an MS in industrial engineering from University of Pittsburgh, and has over 30 years experience in maintenance management and facilities maintenance including over 14 years in energy and environmental management. Marvin kept finding motors running 24/7 when they were only required to run 7-8 hours daily. Also, he observed switches stuck on equipment, sensors not working, building automation systems with operators not trained, data centers using servers that were energy hogs and cold air mixing wrongly with the hot air on the way to the CRAC (computer room air conditioner). He recognized that a maintenance program needed to address equipment using excessive energy. Of the six different maintenance systems, none addressed this energy waste as the primary focus. Marv was excited to present this concept on AEE (Association of Energy Engineers) on-line energy seminars.

Fadi S. Alshakhshir, holds a bachelor degree in mechanical engineering from Jordan University of Science and Technology, and a Masters of Science in Energy from Herriot Watt University. He is a Certified Energy Manager from AEE and attended in early 2016 the ECM seminar that Marv instructs. By the end of 2015, Fadi was already developing a maintenance program that focuses on including energy-related maintenance tasks with regular reliability maintenance plans. He called it Energy Centered Maintenance which became a similar program to what Marv invented. The initial ECM concept was enhanced tremendously by his development of the technical steps necessary to implement ECM and the extension of its application to additional equipment and extending the idea outside of building systems to water supply systems, drainage systems, and fire protection systems. It was apparent the world could benefit if both Marv and Fadi would come together and write a comprehensive ECM book outlining how to implement and sustain this new and beneficial green energy maintenance program.

The title expanded to *Energy Centered Maintenance—A Green Maintenance System*. Why was green maintenance selected? Going green means reducing the overall environmental impact. It also includes compliance with environmental regulations and laws. ECM is an energy efficiency initiative and has as a goal of reducing power consumption thus saving resources, reducing costs, and supporting our environment. ECM implementation is truly greening maintenance. ECM is also an energy efficiency measure and can offset some need for generation and help reduce green house gas emissions.

Acronyms

ARCWT—Actual return chilled water temperature
ASCWT—Actual supply chilled water temperature
DRCWT—Design return chilled water temperature
DSCWT—Design supply chilled water temperature
AHU—Air Handling Unit
BAT—Best Available Technique
BAS—Building Automation System
BMS—Building Management System
CSFs—Critical Success Factors
CCU—Close Control Unit
CRAC—Computer Room Air Conditioner
CUSUM—Cumulative Sum
CMMS—Computerized Maintenance Management System
CMMIS—Computerized Maintenance Management Information System
DPT—Differential Pressure Transmitter
DPS—Differential Pressure Switch
ECMs—Energy Conservation Measures
ECC—Energy Classification Code
ECM—Energy Centered Maintenance
EMS—Environmental Management Systems
EnMS—Energy Management Systems
EnPIs—Energy Performance Indicators
EPEAT—Energy Performance Environmental Attributes Tool
ESPC—Energy Savings Performance Contracts
EUI—Energy Utilization Index
ECI—Energy Cost Index
EPI—Energy Productivity Index
FCU—Fan Coil Unit
HVAC—Heating, Ventilation and Air Conditioning
ISO—International Standards Organization
IT—Information Technology
kWh—Kilowatt Hours
KPIs—Key Performance Indicators
KRIs—Key Result Indicators
LDT—Low Delta T (Low Temperature Difference)
MTBF—Mean Time Between Failures
MTTR—Mean Time To Repair

MCC—Motor Control Center
NGT—Nominal Group Technique
OEE—Overall Equipment Effectiveness
O&M—Operation and Maintenance
OEE—Overall Equipment Efficiency
PAL—For meetings, develop Purpose, an Agenda, and Limit the time per agenda topic.
P-D-C-A—Plan, Do, Check, Act The Deming Wheel
PE—Professional Engineer
PF—Power Factor
PIs—Performance Indicators
PRV—Pressure Reducing Valve
PPM—Planned Preventive Maintenance
PM—Planned Maintenance
PUE—Power Usage Effectiveness
RCA—Root Cause Analysis
QMS—Quality Management System
QVS—Quality Value System Corporation
RCM—Reliability Centered Maintenance
SMART—Specific, Measurable, Actionable, Relevant, and Time-framed.
SWOT—Strength, Weakness, Opportunities, Threats
TPM—Total Productive Maintenance
T&C—Testing and Commissioning
UESC—Utility Energy Services Contract
VFD—Variable Frequency Drive

Chapter 1

Energy Reduction

ENERGY COST

In the United States, around $ 500 billion a year is spent on energy. Energy costs normally represent up to 30% of most corporations operating expenses. The U.S. Green Buildings Council estimates that commercial office buildings use, on the average, over 20 percent more energy than they should, which is a significant dollar loss to industry due primarily to the fact that management does not know where the waste is occurring and how to eliminate or reduce this loss.

The three ways to reduce energy consumption are shown in Figure 1.1 Energy Reductions.

Figure 1.1 Energy Reductions

A common **goal** established by many organizations is to reduce energy consumption by 10% in next two years compared to the energy baseline of last year. The goal drives action and the components contributing look like this: Goal—Strategy—Objectives—Action Plans or Projects.

A **strategy** is simply a plan of how you are going to achieve the goal. Our present strategies shown in Figure 1.1 Energy Strategies are:

1. Implement the low hanging fruit and address identified energy waste.
2. Implement energy conservation program in our organization.
3. Implement energy efficiency measures which include energy efficiency projects and energy centered maintenance.

IDENTIFYING LOW HANGING FRUIT

Low hanging fruit are the most cost-effective actions that can reduce energy use and costs. These measures can be applied immediately such as behavioral changes as switching off unnecessary lights or adjusting set points and time schedules for HVAC systems, or that require little investment such as conducting testing and balancing for some mechanical systems. The simple payback can be immediate, and seen in the future energy bill.

Low hanging fruits are those hundreds of things that are available for an organization to select (pick) and implement at no or little cost but do reduce energy. They are:

• Not Already Implemented
• Easy to Implement
• For My Organization, Low Cost or No Cost
• Can Sell to My Management
• Will Reduce Energy Consumption

Examples are:

1. Establish a compelling energy policy.
2. Implement an energy star procurement policy.
3. Turn off lights & communicate energy conservation plan to all personnel.
4. Unplug appliances and electronics when not in use.
5. Use power chords that turn off when not in use.
6. Verify equipment operational hours and time-schedule.
7. Checking illumination levels and switching off excess lighting.

8. Ensure doors and windows are closed as much as possible to prevent heat loss or infiltration.

9. Check door or windows sealant & insulation performance.

10. Conduct an energy awareness campaign that educates the staff, residents, and tenants about their impact on energy use.

IDENTIFYING ENERGY WASTE

Brainstorming Sessions

Identifying energy waste is an excellent strategy. Once you know where and what the energy waste is, it is possible to develop counter-measures that eliminate or minimize them. There are several methods to do so. Re-commissioning or an energy walkthrough audit is the two most known. However, the organization may have to pay a cost to get these done. Some utilities will do free for their customers. In all most every case, these two methods will result in a cost avoidance or savings well above their cost to accomplish. Two cheaper methods can prove excellent at identifying energy waste; they are management/employee brainstorming sessions and energy walkthroughs.

Management/Employee Brainstorming Sessions

First, the team should develop an energy awareness training. Next, the energy team leader, energy manager, or energy champion should accompany a member of top management to the employee/management brainstorming session consisting of a large department's personnel or several small departments or sections.

The senior management representative gives a short speech mentioning the organization's energy goals, why and purpose of the brainstorming session that the organization is going to engage its entire people in reducing the energy consumption and cost. Next, the energy awareness training is given by the energy rep that came with the top management rep. After the energy awareness training, the energy rep with the help of a scribe will:

1. Write on a whiteboard or a pad on an easel the meeting primary purpose "What Energy Waste is Experienced or Evident in your work area?"

2.   Perform "Silent Generation" by having each person identify three energy waste items in their job areas. For example, computer monitors and CPU are not turning off after being idle, the brightness of computers has not been reduced, not using duplex printing and curtains are over windows not letting light into the room thereby reducing the lumens in the work area.

3.   Go "Round Robin" by having the energy rep go around the room and have each participant offer one of their three suggestions and have it written on white board or pad by the scribe. Continue until all possible ideas have been written.

4.   Discuss each idea, eliminating duplicates, altering some by consolidation, etc. until a final list is obtained.

5.   Normally the ideas are prioritized and selections made. However, in this situation, the list is given to the energy team to do the selecting.

These brainstorming sessions are conducted throughout the organization to get inputs and ideas from all the team's personnel. It gets everyone engaged and gives top management a chance to show their support and commitment. The energy team will consolidate the lists into one main list. The energy team will determine a countermeasure for each idea that will eliminate the waste or at least reduce it. The energy team will track the list until countermeasures have been implemented. Ongoing communications as to progress and results should be provided to the organization's personnel.

Walkthroughs or Energy Audits

Energy walkthroughs are investigations and analysis of facility energy use; it's aimed to identify measures for energy reductions and savings in greenhouse gas emissions. Further, energy walkthrough results in financial benefits by reducing energy consumption. Energy walkthroughs are essential for identifying energy management measures.

To perform an energy walkthrough, several tasks are typically carried out depending on the type of the walkthrough and the size and function of the building. Therefore, an individual energy walkthrough procedure shall be put in place for each facility by itself. Energy walkthrough results in identifying possible energy management measures, and it directs the energy management program to the largest energy use equipment.

The energy team should perform the detailed energy walk-through and prepare a comprehensive report of findings and recommendations inclusive of feasibility study and return on investment calculations. The report should identify a clear projection of the energy consumption reduction and savings subject to this walkthrough audit.

Saving calculations and energy use reduction should include the following:

- Projections of savings.
- Energy efficiency measures.
- Comparisons with baseline data.
- Tariff rates.
- All anticipated costs for energy efficiency measure with its return on investment.
- A precise time bounded plan for implementation of actions.

Energy walkthroughs are inexpensive and can produce excellent ideas on how to reduce energy use and consumption, providing the team members are experienced in doing energy audits and have facility maintenance and engineering experience.

Purpose: To identify energy waste and determine the appropriate fix.

Who? Facilities, engineering, technicians, energy team leader and others that can contribute.

What? Kickoff meeting (optional) walk the building and yard and record anything that uses energy, what it is, the amount of energy used (if possible), whether it can reach a state of excessive energy consumption, what preventative maintenance is being performed now, and other pertinent information.

Walkthrough Focused Areas
Observations
1. Occupancy Sensors

Observe infrequently visited areas and determine whether an occupancy sensor will save energy. Look at restrooms, break rooms, copying or printing areas, mechanical areas, hallways, and other areas.

2. Lights in administrative areas.

Note type such as T-12s, T-8s, & T-5s. Look for areas daylighting can be used and skylights would help. Look at light bulbs and see if they are dirty with film covering them.

3. Building Envelope

Search for leaks in doors and windows. Determine if windows should be glazed, caulked or replaced. Weatherstrip the doors where needed, or replace them.

4. Walls and Roof Insulation

Check the insulation level and determine if more would help.

5. Motors and other equipment except for HVAC

Note each and check the switches and sensors associated with each. Check time-schedule of each equipment and if it runs according to it or continuously running.

6. Data Centers

Look for hot and cold aisles and whether hot air is kept from commingling with the cold air on its return to the computer (CRAC).

7. Security Lights

Check to see if they are adequate and energy friendly.

8. HVAC

Note brand, capacity, date installed, the motors and switches associated with the system, check roof vents and other parts for adequacy and maintenance.

9. Building Automation System (BAS) and metering

See if BAS is outdated. Note where additional metering can help identify potential problem areas.

10. Computers, monitors, imaging equipment, fax machines and other office equipment.

Note: If IT power management is being used. Is the energy equipment environmentally friendly? Is a network being used?

Walkthroughs can also be done along with ECM when the machine is selected to be in ECM or when determining the significant energy users to comply with ISO 50001 Energy Management Systems (EnMS). The walkthrough results will be placed into low hanging fruit, energy conservation, energy efficiency, and energy centered maintenance programs for resolution and energy consumption reduction.

ENERGY CONSERVATION

Energy conservation refers to reducing energy consumption by using less of energy input. Energy conservation is different from efficient energy use, which is using less energy for the same or more output. Driving for less time is an example of energy conservation while driving a vehicle that gets more mileage per gallon is an example of energy efficiency. Turning out the lights when not in use, unplugging appliances or electronics when not in use, making your computer monitors go to sleep after a period of idle time, and using duplex printing when possible are examples of energy conservation items. All organizations wishing to reduce energy conservation should develop and implement an energy conservation program and recognize and reward success. A simple, but effective, energy conservation program is shown in Figure 1.2.

Six C's—A Very Simple Model

Figure 1.2 A Simple, But Effective Energy Conservation Program

Step 1. Create an Energy Conservation Program

Put the energy conservation program into a power point presentation. Include the typical items such as:

1. Switch off the lights when space is not used.
2. Unplug appliances, equipment, and electronics when you leave the room.
3. Purchase energy star equipment, appliances and electronics. It is recommended that all organizations make this a procurement policy.
4. Set computers to turn off after 15-30 minutes of non-use (hibernate or system standby).
5. Set monitors to go to the sleep function after being idle for 15-30 minutes.
6. Turn down the brightness of the computer monitors and televisions.
7. For small refrigerators: Put a bag of ice in the refer. Clean the coils periodically. Un-Plug when on long holidays and vacations.
8. Use power cords that turn off when not in use.
9. When printing or copying use 30% or higher recycled paper.
10. Use duplex printing when feasible.
11. If practical, use network printing instead of everyone having a printer.
12. Use electronic files when possible. Do not keep paper backup unless required by headquarters.
13. Put on your emails: "Do not copy unless necessary. save paper."
14. Report any energy or water problems to facility management.
15. Use a CFL bulb or LED fixture for task lights.
16. Actively participate in alternative work programs.
17. Wear proper clothing when ASHRAE temperature settings are set for cooling and heating. Do not tamper with the thermostats.
18. At work or on campus, take short showers (2 minutes recommended.)
19. Use only the paper towels needed. Organization should use only heavier paper towels so that 4 or 5 do not come out at a time.
20. Check faucets in your area for leaks. Report any leaks to facilities management.
21. Use the stairs for first two floors going up and first three floors coming down.

. Conserve

teps 1-4 help an organization implement energy conservation
results in conserving electricity. To sustain the program, the
e results must regularly be communicated through speeches,
eetings, or on a dashboard. Most importantly, individuals who
uted and made significant contributions are recognized and
ed.

GY EFFICIENCY PROJECTS

he Department of Energy and Climate Change defines energy
cy as "a measure of energy used for delivering a given service.
ving energy efficiency means getting more from the energy that
." <u>Doing the same but with less energy or energy efficiency is
the same mission with less energy</u>.

xamples of energy efficiency projects are:

Jpgrading HVAC.

'utting double paned windows in the facility

eplacing the UPS for a data center.

eplacing and modernizing the cooling tower.

'utting panels in a data center to block the cold air returning to
he inside computer (CRAC-Commuter Room Air Conditioner)
nd mixing with the hot air.

nergy efficiency projects can significantly reduce energy con-
ion. The walkthroughs can help identify energy efficiency proj-
Jext, the payback (the time the savings will pay for the project) is
ted or calculated and projects with a payback of 3 years or less
ost probably be funded by the organization. The government
nsider plans up to 11 years payback to reduce energy consump-
nergy centered maintenance is an energy efficiency measure. An
ization should have a mixture of energy conservation, energy
ncy, and low hanging fruit items implemented in their energy
ion program.

22. Only use a cold wash.
23. Set hot water heaters to 120 degrees.
24. Only use dishwashers for full loads.
25. Use blinds, curtains in all areas to help
26. Keep windows and doors closed when
    is provided.

Step 2. Communicate

Develop an energy conservation prese
your energy champion or manager to all pers
should include the company or organization's
the definition of energy efficiency and energy
energy conservation training including gettir
and do the 26 items above is our simple, but ef
tion program. Experience has shown that impl
conservation items can reduce our energy cor
personnel should receive this energy conservat
a certificate showing they did, and it should
training records.

Step 3. Commit

During the energy conservation training
commit to practicing energy conservation. A
on earth day, April 22 each year, have either i
sheets or put the 26 items on a large sheet a
voluntary commit by signing the large sheet. T
the next year they will support and do energy
help the organization achieve its energy reduct
they do this? Because it is just the right thing
dividual, the organization, and our planet.

Step 4. Changes and Culture

To have a successful energy conservation
have to change their behaviors to participate.
contribute but may not know what to do. The tr
problem. Some who knows what to do- do not c
will help enforce this type of behavior. In Chapt
Centered Behavior leading to an Energy Center
more about achieving the desired behavior.

Step

whicl
positi
staff
contr
rewa

ENEI

efficie
Impr
we u
doing

1.
2.
3.
4.
5.

sum
ects.
com
will
will
tion.
orga
effici
redu

Energy Objectives and Targets with Action Plans

A best practice is to have a cross-functional energy team representing all the major organization's departments. They will need to be trained in energy awareness, process improvement, problem solving and development of objectives, targets, and action plans.

There are three types of goals. The first is one to improve something such as reducing energy consumption, increasing the number of suggestions or ideas, or development of a new procedure. The second purpose is to maintain something you desire not to degrade such as environmental performance. The third type of objective is to determine if something is feasible or not. Feasibility of going on Time In Use Energy, Accepting Demand Response, changing water heater from gas to solar, are examples of this type objective. During the first year, several of the objectives will be developing something such as energy policy, energy plan, energy procedures, development of energy awareness training and others. Then emphasis on reducing energy consumption objectives and targets will be prevalent. Installing occupancy sensors in restrooms, break rooms, mechanical rooms, and other areas with occasional visitors throughout day or night shift. Reducing office paper use and implementing IT power management are good examples.

Objectives of Energy Reduction

The overall purpose of energy reduction is to achieve and maintain optimum energy purchase and utilization throughout different consumer types, such as factories, commercial and residential developments.

The implementation of an energy management process reflects the organizationally responsible behavior in preserving natural resources, reducing the impact on the environment, reducing greenhouse gas emissions, improving air quality and limiting global climate change.

Therefore, the implementation of an Energy Reduction Program will result in the following objectives:

- Enhance energy efficiency continuously by implementing an effective energy management program that supports all operations and achieves customer satisfaction while providing a safe and comfortable environment.

- Developing and maintaining effective monitoring, reporting, and management strategies for wise energy consumption.

- Finding new and better ways to increase returns on investments through research and development and energy saving initiatives.

- Developing interest and dedication to the energy management program from all building's operators, employees, tenants, shareholders, owners, and visitors.

- Reducing operating expenses and increasing asset values by actively and responsibly managing energy consumption.

- Reducing greenhouse gas emissions, mainly CO_2 emission and reducing carbon footprint, caused by energy consumption.

- Complying with regulatory laws & legislations listed by the government.

- Support the growth of renewable energy resources & sustainability commitments.

Characteristics of a Successful Energy Reduction Program
Eight characteristics keep showing up at organizations that have been successful at reducing energy consumption and energy costs. They are:

1. Top management leadership supports, committed and involved in the energy reduction effort and becomes the programs GLUE (Good Leaders Using Energy).

2. Energy reduction is made a corporate priority.

3. Corporate goals are established and communicated.

4. The energy champion or energy manager or both along with their cross functional energy team select challenging strategies that include development of an energy plan and objectives and targets with actions plans.

5. Both key performance indicators (KPIs) and key results indicators (KRIs) are employed and kept current and visible to measure & drive progress and results.

6. Sufficient resources are provided to fund or ensure adequate.

7. An energy centered culture is achieved.

8. Sufficient reviews are conducted to ensure continuous improve-

ment, compliance to legal requirements, and adequate communications provided to keep all stakeholders informed, motivated, and engaged.

For these eight characteristics or success factors, there are best practices that enable them to stand out as crucial to obtaining success. For the first success factor, we have:

- **Top management leadership** supports, committed and involved in the energy reduction effort and becomes the programs GLUE (**G**ood **L**eaders **U**sing **E**nergy). Appoints an energy champion or energy manager who establishes an X-F energy team.
  — Shows commitment in the energy policy. Communicates the policy, goals, success stories, how everyone can support the program.
  — Provides support and resources. monitors energy reduction.
  — Program and leads annual executive review.

- **Energy reduction** is made a corporate priority.
  Top management communicates this in the energy policy. Senior managers give motivating and informative speeches on reducing energy consumption and why it is important. Actions show management commitment.

- **Corporate goals** are established and communicated.
  Energy reduction goal is established. Renewable energy goal(s) are set. Could be multi goals-first two years short-term, middle term 10 years, and neutrality achieved long term goal). Other sustainability goals are set. All goals are communicated to all involved. Green house gas emissions reduction goal(s) is established.

- **The energy champion** or energy manager or both along with their cross functional energy team select challenging strategies that include development of an energy plan and objectives and targets with actions plans.
  — Perform walkthroughs or have an energy audit and re-commissioning.
  — Conduct management/employees brainstorming sessions.

— Appoint functional teams.
— Perform energy research.
— Encourage employee suggestions.
— Develop and train everyone in energy awareness, energy conservation, and energy efficiency.
— Conduct monthly energy team meetings that are effective.
— Use critical success factors (CSFs) to measure progress and drive increased performance.
— Use the energy reduction checklist to identify areas to improve.
— Use both key performance indicators (KPIs) and key results indicators (KRIs). Use the KPIs to drive the KRIs that measure the goals. Always include energy consumption per month, energy intensity, percent energy that is renewable, CSFs scores, and energy costs.
— Use data collection forms for each indicator.
— Design and use dashboards if possible.

• **Sufficient resources** are provided to fund or ensure adequate countermeasures are implemented to achieve the corporate goals.
— Develop and implement an energy conservation program.
— Get everyone to commit.
— Identify energy efficiency projects with excellent payback.
— Implement energy centered maintenance (ECM) in buildings, manufacturing facilities, universities and industrial buildings, including data centers.
— Implement the low hanging fruit items.
— Estimate each as to its contribution and have contributions equal or greater than the goals.

• **An energy centered culture** is achieved. Organization's values and principals identified and taught.
— Energy training provided.
— Committed and caring leadership is apparent.
— Desired behaviors are encouraged and training provided to ensure what is desired is known by participants.

• **Sufficient reviews** are conducted to ensure continuous improvement, compliance to legal requirements, and adequate commu-

nications provided to keep all stakeholders informed, motivated, and engaged.
— Executive or management reviews are held at least annual with the required inputs and outputs.
— The energy team conducts a legal compliance review to assess whether all legal requirements are being achieved.
— Minutes are kept and made available for interested parties to review.

The above is an overview of implementing a successful energy reduction program at any organization, business, university or college. Energy centered maintenance (ECM) is an effective energy efficiency strategy. Its development including history, steps to implement, benefits, examples such as reducing energy consumption in data centers and manufacturing, measuring the efficiency and effectiveness, finding root causes and fixing them, and other pertinent information will be presented in the following chapters.

Energy Centered Maintenance

Energy Centered Maintenance (ECM) is a continuous improvement maintenance regime that combines the physical preventive and predictive maintenance tasks with energy-related maintenance tasks that maintain the operational parameters of the equipment and its efficiency (i.e. motor current, or fan flow rate).

The primary purpose is to reduce energy use by identifying equipment or items that can become energy hogs while still performing their function and prevent that from occurring.

Energy centered maintenance supports the energy reduction program adopted within the facility and helps achieving the projected energy savings.

Why is ECM needed?
• Poor maintenance of energy-using systems, including significant energy users, is one of the leading causes of energy waste in the Federal Government and the private sector.
• Energy losses from motors not turning off when they should, steam, water and air leaks, inoperable controls, and other losses from inadequate maintenance are large.
• Uses energy consumption excess or energy waste as the primary

criterion for determining specific maintenance or repair needs.
* Lack of maintenance tasks in measuring the operational efficiency of the equipment such as motor power consumption and equipment effectiveness.

The walkthroughs or energy audits, the management/employees brainstorming sessions, energy conservation items and energy efficiency projects, implementing low hanging fruit items, and implementing energy centered maintenance are excellent ways to reduce energy waste in your organization, business or university. There are a few administrative moves that will help the energy waste reduction strategy. First, have an energy champion or energy manager to run the day to day energy-related activities and to coordinate with the maintenance team to ensure the maintenance activities are conducted effectively. He or she can benefit by having a cross-functional energy team with representatives from all the main departments including one or two members from facilities and one from engineering. For large organizations, there may be several other teams that either report and inform the energy manager and seek his support and guidance. On a university campus, you can have energy conservation teams for large buildings. Functional teams for major departments identifying waste and developing countermeasures that minimize or eliminate the energy waste that is resulted from lack of proper maintenance.

The ECM planning, designing, and implementing efforts can be accomplished by one or two engineers or by a maintenance leader and a small 4-5 team members.

Steps in Implementing ECM

The following steps should help the energy management and maintenance teams in implementing the ECM policy within the organization.

The details of the ECM model steps are discussed in-depth in next chapters in this book.

Some steps will enable ECM implementation to be accomplished efficiently and effectively.

Step 1. Obtain top management approval and commitment to energy centered maintenance (ECM).
* Make them aware of ECM purpose, objectives & concept.

- Estimate ECM's contributions to the organization's energy reduction plan and reducing green house gas emissions.
- Start developing key contacts in other departments that can help ECM become a reality such as operations, logistics, environmental, facility maintenance, and others.

Step 2. Identify the equipment and systems most likely to use excessive energy.
- Make a list of these systems and equipment, then prioritize them.

Step 3. Determine what systems will be needed to track the ECM activities.
- Consider all systems that are major contributor to energy consumption, and determine which equipment are most energy consuming.

Step 4. Pilot a potential energy hog system.
- Commit to addressing at least one of these troubled energy hog systems for validation of ECM or as a pilot where value can be shown, and proven and baseline information can be developed.
- Begin base-lining/tracking this system.
  — System operations and history.
  — System maintenance and history.
  — System costs, time to service, downtime, resulting over time, OEE, machine efficiency, etc.

Step 5. Determine what ECM tasks need to be conducted.
- Consider required tasks, skill set requirements, tools and equipment, cost effectiveness.

Step 6. Determine what proactive measures should be included in the regular maintenance plan.
- Consider purchasing or enhancing a computerized maintenance management system and commit to its implementation and use, and update it to include the ECM tasks.

Step 7. Purchase diagnostic, volt meters, amp meters, other metering, or monitoring equipment necessary for ECM.
- Be sure not to purchase and equipment or items for inspections if the organization already has them.

Step 8. Achieving maintenance and operational efficiency.
- Understand how to operate this system correctly.
  — Define and complete operator training needs.
- Understand how to maintain this system correctly.
  — Define and complete maintenance training needs and establishing specific inspection procedures, what to look for, and maintenance tasks.

Step 9. Train appropriate personnel in ECM purpose, concept, benefits and how it fits into present maintenance program.
- Train maintenance manager, supervisors, technician leads.
  Who: Maintenance manager, maintenance supervisors, technician leads or foremen, technicians or mechanics.
  What:
  1. ECM purpose, objectives and concept.
  2. Seven types of maintenance—advantages and disadvantages.
  3. Inspections—What to look for, tools needed, data needed.
  4. Equipment identification codes.
  5. Updating PM plans.
  6. Sample problem, cause, effect, & corrective preventative problem.

Step 10. Train technicians or mechanics in ECM procedures.
- Equipment identification.
- Inspections requirements.
- CMMS and data requirements.
- Maintenance, repair, or replacement decisions.

ECM Operating Principles
  Five operating principles guide ECM. They are:
1. Find waste and eliminate it.
2. Perform quality inspections and maintenance or replacement.
3. Be both efficient and effective.
4. The maintenance program management team continually analyzes the maintenance data to identify trends, inefficiencies and to develop strategies for operational and financial improvements. continuous improvement is a goal.
5. ECM addresses and solves all related environmental requirements, concerns or issues.

Some of the five are self-explanatory but not 3, 4, and 5.

Be Both Efficient and Effective

Someone once stated, it is possible to be efficient but not effective or productive but not efficient. Our goal should be that the maintenance technicians are both efficient and effective. For example, operations release Machine XYZ to support to test for excessive energy use on Wednesday morning. They want the machine back on Thursday morning by 10:30 am. The maintenance technician checks the volts and amps and found that both exceeded their nameplate amounts. The machine's bearings were grinding and needed to be replaced. Inventory management only had one-half of the bearings needed and had to order from their central parts store area the remainder. Although they had them on special order, they came in overnight arriving at 8:20 pm. The technician picked them up from the parts delivery person and rushed to the machine. He quickly installed them and had the equipment running smoothly using only nameplate power by 9:50 am. The operations required date was met, so the job had been useful, but was not efficient since the technician had to travel to the machine site more than once and did not have the parts needed. The previous sentence is an example of "Being Effective But Not Efficient." What about being efficient but not effective? Let's say another technician was assigned to do an ECM inspection of machine HWX that was released to maintenance by operations for only two hours, starting at 8:45 am. Operations were nervous about the short interval to do the inspection and possibly some maintenance. The technician did the inspection and found excessive energy was being used. He found the trouble or root cause right away and started the job tasks immediately. The work that had to be accomplished was huge and required more than one technician. They completed the work at 11:00 a.m. within the man-hour estimate or standard but were 15 minutes late to operations thus delaying production 15 minutes of unplanned downtime. The technicians were efficient but not effective.

The maintenance technicians have to consider operations as their customer. The operations required date is their customer's requirement, and they need to meet it. However, in the past, one of the authors, has seen them act like they are each other's enemy since maintenance felt the operators set non-realistic requirements and the operators did not think maintenance planned the job well and took unnecessary breaks or were unproductive sometimes. They must work together as a team respecting each other and honoring each need.

Analysis of the Maintenance Data and Continually Improve

The computerized maintenance management system (CMMS) provides a lot of meaningful data. The number of job plans completed on time. The percent of the time, targeted machines have excessive energy use. The number of times the equipment could be fixed with just maintenance or had to be repaired or replaced. The times needed to do the inspection and the maintenance so historically standards could be developed or compared to the present standards to be sure they are realistic. Data should be trended, and the trends analyze to learn what is happening, what is going good and what should be changed.

Continuous improvement should occur if the plan-do-check-act wheel is followed. Job plans are developed, they are applied by the technicians doing the inspections and maintenance, recorded by CMMS and then provided to the maintenance analyzed to check to see if the job plan was okay or needed change.

ECM addresses and solves all related environmental
requirements, concerns or issues.

Today, both operations and maintenance must be environmentally aware and management the environmental aspects that can impact the environment. Maintenance typically deals with numerous environmental aspects that could impact the environment and the workplace. To name a few: grease, oil, lubricants, cleaning supplies, aerosol cans, batteries, electronic equipment, refrigerants, flammable materials, fuel, bulb disposal, acids, paint, ether glycol, foam, packaging, and compressed gas. Each of these aspects can impact the workplace or the environment. Some of them have legal requirements that must be dealt with to ensure legal compliance. The aspects vary on the risk as to the seriousness of impact, the probability of impact, whether an operational control has been designed to help minimize the impact. They also use and must deal with hazardous materials which must be managed effectively to ensure compliance with federal and state laws. If not, fines can be levied, and the government can close your workplace due to non-compliance environmental problems.

All maintenance technicians or mechanics and their supervisors must be trained on environmental aspects that could impact the workplace. Issues such as lubricants, acids, oil and other items getting into the ground water, what to do in case of an environmental spill, what to do with aerosol cans, how to store flammable materials and other environmental areas. Each supervisor, technician, and mechanic must become an environmental steward and help keep their workplace safe and productive.

ISO 14001 Environmental Management System is one of the best ISO standards written. It provides excellent guidelines and procedures to enable any organization to properly management its aspects in their workplace.

By adopting the ECM Model, the energy and maintenance teams ensures that they are minimizing the impact of equipment operation on the environment. Minimizing the energy waste during equipment operation will reduce the generation of greenhouse gases.

Chapter 2

Different Maintenance Types and The Need for Energy Centered Maintenance

HISTORY OF MAINTENANCE

It is hard to say exactly when maintenance started. During the industrial revolution, some equipment maintenance was accomplished. Facilities maintenance for years was normally emergency maintenance such as a break in the water line, electricity outage in the building, a leak in the roof, or a broken window. Those examples came to be called breakdown or reactive maintenance. Preventative maintenance came in 1951, followed closely by periodic maintenance, and predictive maintenance, all three primary programs. They were defined, procedures were written, maintenance planning and maintenance records became a reality. Non-emergency tasks such as painting, lubricating, replacing bearings, replacing burned-out light bulbs, repairing door locks, caulking around windows, replacing filters, etc. became part of the maintenance of all the main buildings since management now believed that was the most cost effective way of preserving their investment. Predictive maintenance was valued in the manufacturing facilities in that machine efficiency and personnel productivity benefitted from this new predictive capability.

In 1978, Total Productive Maintenance (TPM), invented by the Japanese as part of their Total Quality Improvement Effort was introduced by Toyota in a widely acclaimed book on Toyota's achievements in quality improvement. The book is "Toyota Production System—An Integrated Approach to Just-In-Time." TPM got the employees engaged and followed the simple concept "Everything has a Place and Everything in its place." Cleaning the workplace was one of the first TPM's requirement. The concept also uses the six S's technique to get everything in its place, standardized and sustained. About this time, a new concept called Reliability Centered Maintenance (RCM) came along and said all equipment are not equal. Sometimes we need to perform much maintenance on some equipment and let others run to they failed. RCM included using statistics and determining root causes, thereby limited its application.

In 2012, an often used strategic goal or objective in most large businesses, universities, and organizations was to reduce energy consumption which saved money and contribute to reducing greenhouse gas emissions. Energy as an aspect in ISO 14001 Environmental Management System (EMS) was not receiving sufficient attention, so ISO 50001 Energy Management System (EnMS) was created. Maintenance managers felt energy and maintenance were related but had not defined how. Motors running when they should not, machines running while using excessive electricity, servers using excessive energy, and other similar happenings resulted in energy centered maintenance being born in 2012. Although seminars on ECM have been given since 2013, no book has been written. This book is the first book written on ECM to enable energy savings from maintenance practices to be realized.

THE MAINTENANCE TYPES

There are seven recognized maintenance types counting energy centered maintenance. They are:
1. Breakdown or reactive maintenance (before 1950, manufacturing revolution)
2. Preventative maintenance (1951)
3. Periodic maintenance (1951)
4. Predictive maintenance (around 1951)
5. Total productive maintenance (1951 origin, the 1980s in the USA)
6. Reliability centered maintenance (1960s origin, 1978 became known)
7. Energy centered maintenance (2012)

The reference "htpps://www1.eere.energy.gov/femp/pdfsOM_5.pdf" available on the internet will be used since it best describes the first six maintenance types.

BREAKDOWN OR REACTIVE MAINTENANCE

Basic Philosophy
* Machinery runs until it fails.
* Repair or replace damaged equipment only when problems occur including failure.

Reactive maintenance is often thought of as a "run it 'til it breaks" maintenance strategy.

No maintenance tasks are taken to maintain the equipment as the manufacturer originally intended.

Breakdown or reactive maintenance can be defined as activities are also known as "run-to-failure." Using this approach, maintenance will be performed just when the asset's deterioration causes a functional failure. Reactive maintenance is ideally used when the failure does not significantly affect operation, production, or generate any financial losses other than repair costs if the dollar impact is less than the cost of preventing the failure.

Advantages
* Low cost.
* Less maintenance personnel

Disadvantages
* Additional cost due to downtime of equipment that is not expected.
* Labor Cost could increase if overtime is necessary.
* The cost of fixing or replacing equipment could increase.
* Could have secondary equipment or process damage from equipment failure.
* Inefficient use of maintenance personnel.

PREVENTIVE MAINTENANCE

Basic philosophy
Maintenance tasks are scheduled for predetermined time intervals.

The objective is to repair or replace any damaged equipment before any problems occur.

Preventive maintenance is defined as activities performed on a schedule that is detected via inspection. By performing the tasks, degradation of a component or system is prevented.

The proven benefit of preventive maintenance is that it provides control of maintenance beyond the reactive level to prevent failures. Managing preventive maintenance does minimize possible failures, yet it does have a risk. The maintenance intervals chosen for the "life" of the asset may not be statistically based. Therefore, the likelihood of over-maintaining or under-maintaining the equipment or facility could be very high.

Maintenance activities generated from preventive maintenance and predictive maintenance are called corrective maintenance. These activities should be planned and scheduled in advance of a failure.

Advantages
- Cost effective in all most all organizations (12 to 18 % Cost Savings over Reactive Maintenance (htpps://www1.eere.energy.gov/femp/pdfsOM_5.pdf)
- Flexible
- May achieve some energy savings.
- Reduces equipment failures.

Disadvantages
- Catastrophic failures can occur.
- Labor cost is very high
- Could include the performance of some maintenance tasks that may not need to be performed.

PREDICTIVE MAINTENANCE (CAN BE EITHER CONDITION-BASED OR TIME-BASED MAINTENANCE)

Basic Philosophy
- Schedule maintenance activities when determined that either mechanical or operational conditions warrant.
- Repair or replace damaged equipment before any obvious problems occur.

Predictive maintenance utilities vibration analysis helps predict equipment condition.

Measurements detect the onset of a system degradation, thereby enabling causal stressors to be eliminated or controlled before any significant deterioration occurs in the component's physical state. Results indicate current and future functionality.

The technical basis for predictive maintenance is that most in danger of failing assets provide sufficient warning of the fact that they are in the process of failing during operation. Predictive maintenance replaces preventive maintenance only when the equipment requires advanced warning functional failure, enabling department's resources to be efficiently planned and used.

Advantages
• Increased asset life cycle.
• Allows for the asset to be corrected. Otherwise, the component or asset would fail.
• Decrease in equipment downtime.
• Reduction in parts and labor costs
• Improved environmental safety.
• Improved workplace safety and employee morale.
• Possibly some energy savings could be realized.

It is estimated that 8% to 12% cost savings over preventive maintenance program could be realized (https://www1.eere.energy.gov/femp/pdfs/OM_5.pdf.)

Reduction in maintenance costs: 25% to 30% (https://www1.eere.energy.gov/femp/pdfs/OM_5.pdf

Elimination of breakdowns: 70% to 75% (https://www1.eere.energy.gov/femp/pdfs/OM_5.pdf)

Reduction in downtime: 35% to 45% (https://www1.eere.energy.gov/femp/pdfs/OM_5.pdf)

An increase in production: 20% to 25% (https://www1.eere.energy.gov/femp/pdfs/OM_5.pdf)

Disadvantages
A significant investment in diagnostic and vibration equipment and analysis.

Increased cost in maintenance personnel. Predictive maintenance is different from preventive maintenance in that the actions are determined by the actual condition of the machine or equipment rather than on some preset schedule based on time between maintenance.

Remember that preventive maintenance is time-based. Actions such as changing oil are based on time, like calendar time or equipment run time. For example, most people change their oil in their cars every 3,000 to 5,000 miles. "https://www1.eere.energy.gov/femp/pdfs/OM_5.pdf"

RELIABILITY CENTERED MAINTENANCE

Basic philosophy
A structured/logic-based process used to develop complete system and equipment maintenance programs that will provide the highest level of equipment reliability at the least cost.

Reliability centered maintenance (RCM) utilizes predictive/preventive maintenance techniques and adds root-cause failure analysis to detect and pinpoint the precise problems. RCM includes advanced installation and repair techniques, which could require equipment to be totally redesigned or modified to prevent or eliminate breakdowns from occurring.

RCM believes that all equipment in a building is not of same importance or priority to either the mission, process or facility.

RCM does recognize that equipment design and operation differs and different assets will have a higher probability of undergoing failures than others. RCM will enable an organization to more closely match their resources to maintenance needs while improving reliability and decreasing cost.

Advantages
- Lower cost can be achieved by eliminating unnecessary maintenance or overhauls.
- Reduced probability of sudden equipment failures.
- Enables technicians and mechanics to focus maintenance activities on critical components.
- Increased component reliability.
- Includes root-cause analysis.
- Improves reliability and availability of the machines and other equipment.

- Reduces machines downtime
- Reduces maintenance costs

Disadvantages
- Can have significant startup costs, extensive training, etc.

Not easy to sell to management since it has complicated cause analysis and other evaluations as part of its implementation methodology.

RCM Maintenance Priority
1   Emergency Life, health, safety risk-mission criticality
2   Urgent Continuous operation of facility at risk
3   Priority Mission support/project deadlines
4   Routine Prioritized: first come/first served
5   Discretionary Desired but not essential
6   Deferred Accomplished only when resources allow (htpps://www1.eere.energy.gov/femp/pdfsOM_5.pdf)

TOTAL PRODUCTIVE MAINTENANCE (TPM)

Basic Philosophy
Equipment maintenance including autonomous, planned, and quality maintenance.

Includes training and education and the application of the 5 S's—sort, straighten, shine, standardized, sustain.

The Japanese total productive maintenance (TPM) is a system of maintaining and improving the productivity and efficiency of production and quality systems through the machines, equipment, processes, and employees practices adding business value to an organization.

TPM (Total Productive Maintenance) is a people-involvement approach to equipment maintenance that strives to achieve perfect production: Perfect production means:
- No Breakdowns
- No Slow Running
- No Defects to Parts or Equipment
- Maintaining a Safe and Healthy Work Environment
- No Accidents or Mishaps

TPM uses proactive and preventative maintenance to increase the operational efficiency of equipment. It places a high emphasis on empowering operators to help maintain their equipment.

Therefore, the implementation of a TPM program provides a shared responsibility for equipment that encourages greater ownership and involvement by plant floor workers. In most cases, this can be very effective in improving productivity (increasing machine availability, reducing cycle times, and eliminating defects).

Advantages
- Gets Production & Maintenance Working Together
- Gets Operators Involved
- Uses 5 S's
- Has Focus

Disadvantages
- Training Costs

ENERGY CENTERED MAINTENANCE

Basic Philosophy
1. Find it-fix it
2. Repair or replace when equipment is pulling excessive amps or volts
3. Replace equipment that is an energy hog
4. Operate equipment in a shift only for the time needed
5. Maintain the equipment to ensure operational performance of the machine is functioning efficiently as intended

Energy centered maintenance is defined as a continuous improvement maintenance regime that combines the physical preventive and predictive maintenance tasks with energy-related maintenance tasks that maintain the operational parameters of the equipment and its efficiency (i.e. motor current, or fan flow rate).

The energy centered maintenance model applies to all critical energy equipment to extend its life, maintain equipment efficiency, prevent excess energy use and reduce energy waste.

ECM—Locate and inspect—Test for the energy waste-determine

course of action-repair? Perform maintenance? Replace? Let run to failure? Put into PM program? Put into PM program but as an ECM item? Power quality?

If it meets the ECM test of using excessive energy, develop information for ECM and place in program. Develop a short fact sheet that explains what the function is, how the item can use excessive or waste energy and the cause that can make this happened.

Advantages
- Saves significant energy
- Keeps equipment from failure
- Prolongs equipment life cycle
- Ensures equipment is operating according to intended design function
- Increases energy efficiency of the equipment by achieving a low operation and maintenance cost.
- Can be integrated easily with planned preventive maintenance
- Identifies improvements which can be made to increase equipment's efficiency
- Disadvantages
- Have to find the equipment with the condition

Maintenance Strategy
An effective maintenance approach results from a well-organized and carefully executed effort by the maintenance management team. They identify and define the maintenance tasking, and use established standard failure codes of the proactive approach. The maintenance staff takes input from several sources, including facility/account management, building operations, and the maintenance staff.

A critical system does not imply all maintenance tasks on it prevent failure. Each facility, system, or piece of equipment should be examined/classified by its impact on building operation to determine the criticality of the equipment operation on the plant's business. To determine the maintenance approach, measure the facility, systems, and equipment against the following criteria:
- High repair/replacement costs
- High energy consumption/conservation
- Effect on the facility value
- The method for identifying the maintenance approach by mainte-

nance management is summarized in the following steps:

— Consider the facility goals and requirements for system operation.

— Examine the facility and identify assets whose proper performance is necessary to meet organization business and maintenance requirements.

— Determine the impact (of the functional failure of each facility asset and equipment.

— Determine what predictive or preventive tasking will mitigate potential failures.

If the cost of such maintenance actions is less than the impact of the functional failure, assign the appropriate predictive and preventive maintenance activity on the maintenance schedule.

The maintenance strategy involves tasks that can be applied to prolong the useful life of equipment and prevent/avoid premature failures.

Selecting the appropriate maintenance approach includes the following:

• Consider the criticality of the equipment on facility operation and determine what kind of maintenance approach is best, such as run to failure, preventive maintenance, predictive maintenance, total productive maintenance and reliability centered maintenance.

• Consider the variety of problems (failures) that may develop in equipment.

• If preventive maintenance approach is not adequate to detect the variety of maintenance problems, use predictive maintenance or reliability centered maintenance concepts.

• If predictive maintenance or reliability centered maintenance does not adequately apply or is cost prohibitive, use preventive maintenance. Develop inspection tasks to reveal the failures not adequately covered by predictive maintenance and reliability centered maintenance.

• Decide the combination of approaches (Predictive/Proactive) to develop a total productive maintenance approach, then determine the frequency of the particular task. Usually, a combination of concepts provides the required coverage to assure reliable performance.

• Development of maintenance strategy will help the maintenance management team to:

— Concentrating maintenance resources where they will do the best.
— Performing technically efficient and cost-effective maintenance
— Eliminating unnecessary and ineffective maintenance tasks.
— Developing a documented basis for the maintenance program.

DETECTION OF POTENTIAL FAILURES

The detection of possible future failures in the maintenance strategy yields an important means of improving the overall maintenance program. Early detection of future problems can prevent a critical level of deterioration or failure, so the maintenance organization can use lead time to deliberately plan repair or improvement work, then carry it out using resources more efficiently.

The maintenance strategy is based on the fact that many failures do not occur instantaneously, but develop over time. If evidence a failure is occurring can be found, it may be possible to take action to prevent the failure and avoid the consequences.

The detection of potential failures approach focuses on the failure interval allowing the maintenance organization the opportunity to use the time to deliberately plan repair or improvement work, then carry it out using resources more efficiently.

The principal aim of the maintenance approach is to uncover problems or manageable effects before they reach the crisis stage of equipment failure or breakdown. The sooner the problems or effects are found, the greater opportunity for planning, gathering materials, coordinating outages, estimating, and allocating resources.

Chapter 3

The Energy Centered Maintenance Origin and Model

ORIGIN OF ECM

Years ago, after giving a speech on maintenance at an industrial engineering conference, I was asked: "Is there any connection with maintenance and saving energy?" I had not been asked this question before. My answer is yes. If we paint a room white, it will take fewer lumens to serve the place. If we change a filter in an air conditioner on time, the air conditioner will not have to operate with higher airflow resistance, thus saving electricity. That question in the last 15 years has often been asked. Organizations that provide facility and equipment maintenance would like to advertise or be able to tell their client they save energy while providing their maintenance. They were hesitant until ECM came along. ECM shows a direct relationship between reducing energy consumption and performing maintenance tasks. They knew that.

- Poor maintenance of energy-using systems, including significant energy users, is one of the leading causes of energy waste in the federal government and the private sector.
- Energy losses from motors not turning off when they should, steam, water and air leaks, inoperable controls, and other losses from inadequate maintenance are large.
- Uses energy consumption excess or energy waste as the primary criterion for determining specific maintenance or repair needs.
- Lack of maintenance tasks in measuring the operational efficiency of the equipment such as motor power consumption and equipment effectiveness.

ECM uses energy consumption excess or energy waste as the primary criterion for determining specific maintenance or repair needs.

energy centered maintenance (ECM) was originated in 2012 when Marvin Howell kept finding motors running 24/7 when they were only required to run 7-8 hours daily. Also, he observed switches stuck on equipment, sensors not working, building automation systems with operators not trained, data centers using servers that were energy hogs and cold air mixing wrongly with the hot air on the way to the CRAC (computer room air conditioner). He discovered that a maintenance program needed to address equipment using excessive energy. Of the seven different maintenance systems, none addressed this energy waste as the primary focus. Figure 3.1 Maintenance and Energy Relationship shows the relationship. Marv was excited to present this new concept on AEE (Association of Energy Engineers) On-Line Energy Seminars.

Figure 3.1 Maintenance and Energy Relationship

THE MODEL—ITS AIM AND DESIGN

In 2015, Fadi Alshakhshir designed a detailed process of implementation of the energy centered maintenance model (ECM). The aim of the model is to introduce new maintenance tasks to all energy critical equipment, to ensure it is functioning efficiently, and ability to deliver its intended function. The details of the process are explained in this book.

ECM model is a continuous improvement maintenance regime that combines the physical preventive and predictive maintenance tasks with energy-related maintenance tasks that maintain the operational parameters of the equipment and its efficiency (i.e. motor current, or fan flow rate). ECM along with reliability centered maintenance (RCM) aim to improve the equipment reliability and energy efficiency.

The energy centered maintenance model applies to all energy critical equipment to extend its life, maintain equipment efficiency, prevent excess energy use and reduce energy waste.

The two most important ECM strategies are:

- Preventive maintenance (time-based, usage based).
- Predictive maintenance (condition-based).

The character of the facility and equipment type determines which approaches are most effective; some combination of preventive and predictive maintenance is required to assure optimum energy efficiency.

The strategy for selecting the appropriate preventive or predictive energy centered maintenance approach involves the following:

- Consider the variety of deficiencies that may develop in equipment.
- If preventive maintenance is not adequate to detect these deficiencies, use predictive maintenance. One or a combination of several predictive technologies may be required.
- If predictive maintenance does not adequately apply or is cost prohibitive, use preventive maintenance. Develop inspection tasks to reveal the deficiencies not adequately covered by predictive maintenance.
- Decide the combination of approaches (predictive/proactive) then determine the frequency of the particular task.

ECM provides the basis for identifying multiple low or no cost operation and maintenance practices that reduce the use of energy and enhance the efficiency of the equipment. ECM works on the concept of returning the equipment to its original operational parameters (as originally commissioned) which result in improving its energy efficiency and reduce its energy use.

Practical maintenance activities and measures that require very minimal cost or that may not require any investment can be useful measures of the energy centered maintenance program and the operation and maintenance (O&M) regime. Energy reduction can be achieved in numerous items of equipment which results in the overall decrease of building's energy consumption.

Table 3.1 illustrates cost advantages of the two different maintenance approaches in the energy centered maintenance model.

Different studies and experiments have shown that the energy consumption of equipment can be optimized and reduced by 5-15% without a significant investment (PECI 1999). Acknowledging that significant energy reduction can be achieved by related performance operation, and maintenance measures are the main reason why the energy centered maintenance model has been developed.

Energy centered maintenance model is a unique maintenance program that focuses on energy-related equipment such as air handling units, electrical motors, pumps, etc. Those types of equipment should be identified based on their energy consumption, and information should be collected from the testing and commissioning data (T&C) to compare the equipment's current behavior with its original commissioning data.

Well-functioning equipment is safe for the operation of the facility. When equipment is properly maintained its efficiency remains high, and its operating performance remains acceptable across its lifespan thereby minimizing the risk of any deficiencies that could increase the energy consumption of the building.

The proper execution of energy centered maintenance model should be delivered via planned job plans that are defined by the equipment's preventive and predictive maintenance needs. Equipment efficiency is significantly increased when tasks are accomplished by standard proactive programs. Job plans should be designed for each piece of equipment based on relevant factors such as the maintenance tasks, frequency, tools, duration and craft type. The cycle of developing and implementing the job plans is the primary step in an energy-centered maintenance process.

The job plans should be determined, scheduled and updated on the organization's existing CMMS system and existing maintenance reliability program to allow a proper scheduling and monitoring. However, it is not necessarily that the ECM program frequency to follow the regular PM program for cost effectiveness.

The energy centered maintenance model is a process based methodology used to analyze and continuously improve assets and equipment maintenance and energy efficiency. This approach ensures that informed energy efficiency measures are made that will significantly

Table 3.1 Cost Advantage of PM and PDM

Maintenance Type	Frequency	Risk of Deficiency	Cost of Maintenance	Cost of Parts
Preventive Maintenance	Time-Based or Operating Cycles	Low – Scheduled work	Low – Minimum Shutdown Costs Resulting From Time-Based Frequency	Low – Controlled and scheduled frequencies permit fewer inventory parts
Predictive Maintenance	Condition-Based	Lowest – Deficiency avoided through detection	Lowest – Condition Is determined while asset is operating	Lowest – Lead time permits planning of material requirements

impact the energy efficiency of the facility in a cost effective maintenance model.

The model is a seven-step process that involves identifying the
type of equipment and measuring its current performance and improving it (Figure 3.2), the steps are:

- Step 1.0: Identify energy critical equipment.
- Step 2.0: Collect data and define baseline performance.
- Step 3.1: Identify energy centered maintenance inspection & fre-
  quency.
- Step 3.2: Identify craft, tool, and duration.
- Step 4.0: Measure equipment's current performance and compare
  to baseline.
- Step 5.1: Identify root cause.
- Step 5.2: Identify corrective action and check cost effectiveness.
- Step 5.3: Restore equipment efficiency.
- Step 6.0: Update CMMS and plan next inspection.

OBJECTIVES OF ECM

The purpose of energy centered maintenance model (ECM) is to
develop a proactive maintenance approach focused on the analysis of
equipment operational parameters to ensure it is functioning by its design intent. The model will help to reduce the energy consumption of
the equipment and improve its efficiency.

The ECM Process's objective is to increase the energy efficiency of
the equipment in a cost effective manner using proven maintenance assessments and identifying maintenance related tasks that measure and
improve the current operational behavior of the equipment.

Unlike reliability maintenance, the energy centered maintenance
model does not intend to enhance equipment reliability or to prevent
failures; the aim is to create maintenance tasks that prevent energy
waste during equipment operation and to ensure it is delivering the
intended function.

Development and implementation of ECM model also have the
following objectives:

- To provide educational practice of how energy consumption is
  related to maintenance.
- Improving maintenance regime to focus on the operational condi-

Figure 3.2: ECM Model

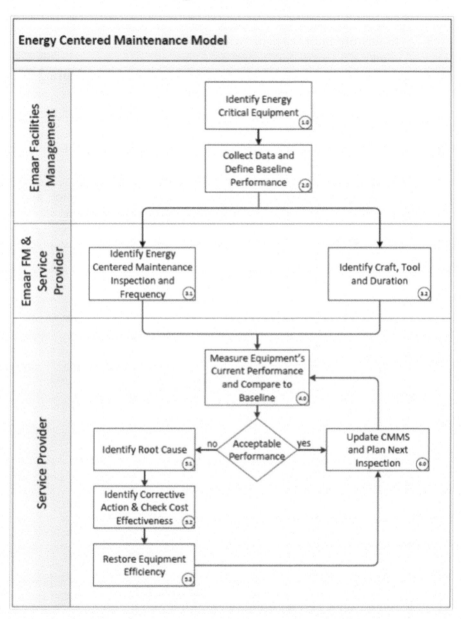

    tion of the equipment.

- To identify any change in the equipment performance compared to testing and commissioning data.
- To identify improvements which can be made to increase equipment's efficiency.
- Optimizing the energy consumption of the equipment during its operation.
- Increase energy efficiency of the equipment through a low operation and maintenance cost.
- Ensure that equipment delivers the expected operational parameters as per the design intent.
- Reduce energy consumption of the facility.
- Reducing greenhouse gas emissions, mainly CO_2 emission and reducing carbon footprint, caused by energy consumption.

Chapter 4

ECM Process —
Equipment Identification

STEP 1. EQUIPMENT IDENTIFICATION

The first step in developing an energy centered maintenance strategy is the identification of all energy related systems (i.e. mechanical system) and all energy critical equipment (i.e. pumps) that should be included in the maintenance strategy of the facility.

The energy related systems and their components differ from one building to another depending on the type of the building, for example, escalators might be available in malls but not necessarily in residential buildings.

Equipment identification defines the energy critical equipment that should be considered for energy centered maintenance approach. Energy criticality is determined based on two main factors:
• Building systems.
• Equipment energy classification code.

Those factors are focusing on identifying high energy consumption systems as well as main energy consuming equipment. Each equipment should be assigned with energy classification code based on the amount of energy consumed by the equipment and its operational hours, which in its turn decides if the equipment is feasible to be included in the energy centered maintenance model or not.

LIST OF ENERGY RELATED SYSTEMS

The first step is to list down all the systems and mechanical or electrical equipment that are energy related within the facility.

The following list of systems and components focus on energy centered maintenance requirements, the equipment which is part of those systems are critical for the energy consumption of the facility:

I. Mechanical Systems:
  1- Heating, Ventilation, and Air Conditioning System:
    Air Handling Units.
    Fan Coil Units.
    Energy Recovery Units.
    Boilers.
    Pumps.
    Close Control Units.
    Fans.
    Cooling Towers.
    Air Cooled Chiller.
    Water Cooled Chillers.
    Heat Exchanger.
    Direct Expansion Air Conditioning Units.

  2- Water Supply System:
    Pumps.
    Heat Exchangers.
    PRV Stations.
    Boilers.

  3- Drainage System
    Sump Pumps (Sewage).

  4- Storm Water Management System.
    Rain Water Pumps.

  5- Building Transportation System
    Elevators.
    Travelators.
    Escalators.

II. Fire Fighting Systems
  Fire Pumps.

III. Electrical Systems:
  Motor Control Centers.
  Variable Frequency Drive (VFD).

IV. Building Management System
    2 Way - 3 Way Valve Functionality.
    Differential Pressure Switch—DPS.
    Differential Pressure Transmitter—DPS.
    Airflow Meters.
    Velocity Meters.
    On Coil Temperature & Humidity Sensors.
    Off Coil Temperature & Humidity Sensors.
    Space Temperature and Humidity Sensors.
    Thermostat Functionality.
    Control logic for all equipment

Each facility shall identify all systems and equipment applicable to their building, the technical information about the equipment can be found in asset registers, equipment schedules, O&M manuals, As-built drawings, etc. This information will be used to define the baseline performance of the energy related equipment and to design a maintenance checklist and plan as part of the preventive and predictive energy centered maintenance strategy.

ENERGY CLASSIFICATION CODE

Energy classification is defined as the process of identifying the equipment's energy impact and criticality to the facility operation. In-depth analysis must be performed for significantly energy consuming equipment before starting energy centered maintenance model and efficiency improvement. Each system should be examined and classified by its impact on the facility operation and consumption, taking into consideration its impact on customers comfort, regulatory requirements, consumption level and environmental impact.

The Energy Classification Code (ECC) reflects the criticality of equipment on building energy use; a high energy classification code indicates high energy consumption equipment. The importance of assigning an energy classification code for each equipment is essential to identify which equipment will be part of the energy centered maintenance model. See Table 4.1.

The ECC has a scale of 1-5, where 5 is high energy critical equipment, and 1 is low energy critical equipment. Thus, the equipment

Table 4.1 Energy Classification Code

Energy Classification	Energy Impact	Description	Examples
High **4 or 5**	Large Impact Energy users	Systems with the following characteristics must be considered highly critical: • Operating Profile: Continuous Running. • Energy Capacity: High Capacity.	i.e. Chilled Water Pumps, AHUs motors, Lighting, etc. - Efficiency loss on these systems will result in high associated energy costs. - Must be continuously running.
Medium **3**	Medium Impact Energy users	Systems with the following characteristics must be considered critical: • Operating Profile: Continuous Running. • Energy Capacity: Low Capacity.	i.e. Toilet Exhaust Fans, Lighting, etc. - Efficiency loss on these systems will have a medium impact on facility energy cost. - Might be running on time-schedule
Low **1 & 2**	Low Impact Energy users	Systems with the following characteristics must be considered noncritical: • Operating Profile: Non-continuous Running. • Energy Capacity: Vary in Capacity.	i.e. Fire Pumps, Stair Case Pressurisation Fans, Emergency Lighting. - Might be high or low energy consumers - Operates only in case of emergency

Energy Classification Code

which results with ECC of 5, 4 or 3 shall be part of energy centered maintenance model, while equipment with ECC of 1 and two are considered of low energy impact and hence it is not part of the maintenance strategy. Equipment with ECC codes 5, 4 and 3 are the ones that will be part of energy centered maintenance plan, and detailed data about the equipment should be collected to start the maintenance planning.

To determine what ECC should be given for certain equipment, mathematical calculations should be made based on two measurable parameters, which are the operating profiles (run time and continuity of equipment operation in hours/day), and energy capacity (high/low) based on the connected load of the equipment.

The following steps explain the process of assigning an energy classification code for building equipment by using the following mathematical equation, graphical scale, and ECC table:

1- List down all energy related equipment that is serving the facility.
2- Identify connected electrical load for each equipment (kW).
3- Determine total daily working hours for each equipment.
4- Calculate daily operating load (electrical consumption) of each equipment (kWh).
5- Calculate daily operating load (electrical consumption) of all equipment (summation of kWh).
6- Calculate each equipment operating weight percentage to total equipment operating load using the following equation (%).
7- Assign energy classification code for each equipment based calculated equipment operating load percentage by using the scale in Figure 4.1.

Example of Calculating and Assigning Energy Classification Code

Consider a facility that is served by multiple mechanical types of equipment, the maintenance personnel and energy managers should be able to calculate the energy classification code for each piece of equipment to decide whether this equipment should be part of energy centered maintenance model or not.

The facility is serviced by the following mechanical equipment:
• Three chilled water pumps
• Three air handling units
• 25 fan coil units
• Three elevators

Single Equipment Load Percentage (%) = Qty x Operating hours day x Connected Power kW Full Day

Load of all equipment kWh day

Figure 4.1 Energy Classification Code Scale

- Three kitchen extract fans
- Three smoke extract fans
- Six toilets extract fans
- Two ejector pumps
- One firefighting pump

Following the steps described above, the energy managers and maintenance personnel should be defined the connected load for each piece of equipment (kW), and the total daily operating hours (hr) so that they can calculate the operating load (kWh) as described in the Table 4.2.

Table 4.2 shows that the total operating load for all equipment is 9154.0 kWh per day which is required to calculate single equipment load percentage to assign the ECC code. For example, for chilled water pumps, the parameters are:

- Connected load for each pump is 50.0 kW
- Operating hours per day for each pump is 24 hours
- Total operating load for all equipment is 9154.0 kWh per day.

Single Equipment Load Percentage % = Qty x Operating hours hrs day x Connected Power kW Full Day Load of all equipment kWh day

Single Equipment Load Percentage % = Qty x Operating hours hrs day x Connected Power kW Full Day Load of all equipment kWh day.

Considering chilled water pumps load percentage is 39.33 %, and by refereeing to the scale, the energy classification code that should be assigned to the pump is 5. (See Figure 4.1)

Considering the same calculation criteria for each equipment type, the ECC for all equipment are shown in Table 4-3.

As can be seen, only equipment with ECC code of 3, 4 and five should be part of the energy centered maintenance model. Equipment with ECC of 1 & 2 don't contribute much to the total energy consumption of the facility and may be not cost effective to include it in the model.

Firefighting pumps remain an ideal example of low energy impact equipment, although their connected load is very high (120 kW in this example), but it works only in the unlikely case of fire. Therefore, it has no significant impact on the energy consumption of the facility.

Table 4.2 List of equipment and its related electrical consumption

List of Equipment and its related electrical consumption

Equipment Type	Quantity	Connected Load kW	Operating Hours per Day (hr)	Operating Load, Consumption for each equipment (kW.hr)	Operating Load, Consumption for all equipment (kW.hr)
Chilled water pumps	3	50	24	1200	3600
Air handling units	3	15	18	270	810
Fan coil units	25	1.5	18	27	675
Elevators	3	75	12	900	2700
Kitchen extract fans	3	15	16	240	720
Kitchen Supply Fan	3	10	16	160	480
Smoke extract fans	3	30	0.5	15	45
Ejector pumps	2	2	16	32	64
Firefighting pump	1	120	0.5	60	60
Total operating load (electrical consumption for all equipment)					9154

Table 4.3. Calculated Energy Classification Code

Equipment Type	Operating Load, Consumption for each equipment (kW.hr)	Operating Load, Consumption for all equipment (kW.hr)	Load % for each equipment	Energy Classification Code
Chilled water pumps	1200	3600	39.33%	5
Air handling units	270	810	8.85%	4
Fan coil units	27	675	7.37%	4
Elevators	900	2700	29.50%	5
Kitchen extract fans	240	720	7.87%	4
Kitchen Supply Fan	160	480	5.24%	3
Smoke extract fans	15	45	0.49%	2
Ejector pumps	32	64	0.70%	1
Firefighting pump	60	60	0.66%	1

Chapter 5

ECM Process —
Data Collection

STEP 2. DATA COLLECTION AND
EQUIPMENT OPERATIONAL BASELINE

All equipment assigned with ECC codes of 5, 4 and three should go through data collection process which must be performed before proceeding further. Data collection is a process of gathering and measuring the required variables and information to support the energy maintenance tasks in the energy centered maintenance model.

Data collection is necessary to define the baseline for the equipment operational parameters, it focuses on obtaining the information about the design of the equipment, and the testing and commissioning parameters which define the base operational behavior of the equipment.

Data collection process is needed as it ensures that data gathered both determined and accurate and that subsequent decisions based on the findings are valid. The process provides both a baseline from which to measure and in particular cases a target on what to improve.

TYPES OF DATA

The energy centered maintenance model focuses on the operational parameters of the equipment. Thus information about equipment's profile and efficiency are essential for the execution of the model.

The type of data that need to be collected differ based on equipment type; data can be found in the testing and commissioning records, O&M manuals, as-built drawings, etc.

The following list of data is a sample of what need to be collected, however, on case by case basis; the FM team shall design this part based on the type of operating systems in their facilities:

1. Heating, Ventilation, and Air Conditioning System:
    - Air handling units:
      Fan airflow rate (m³/hr).
      Motor power (amps, voltage).
      Cooling coil pressure drop (kPa).
      Cooling coil performance
      VFD data
      Control valve control logic
      On coil/off coil temperatures (°C).
      Chilled water Delta T (°C).

    - Fan coil units.
      Fan airflow rate (m³/hr).
      Motor power (amps, voltage).
      Cooling coil pressure drop (kPa).
      Cooling coil performance
      VFD data
      Control valve control logic
      On coil/off coil temperatures (°C).
      Chilled water Delta T (°C).

    - Energy recovery units.
      Fan airflow rate (m³/hr).
      Motor power (amps, voltage).
      Effectiveness (on/off air temperature).
      Cooling coil performance (pressure drop, temperature)—
          Where applicable

    - Boilers.
      Boiler efficiency.
      Electric heater power (kW)—for electric powered boilers.
      Fuel consumption efficiency—for fuel powered boilers.
      In/out water characteristics.

    - Pumps.
      Pump water flow rate (lps).
      Motor power (amps, voltage).
      Pump head (kPa)

- Close control units.
  Fan airflow rate (m³/hr).
  Motor power (amps, voltage).
  Cooling coil pressure drop (kPa).
  VFD data.
  On coil/off coil temperatures (°C).
  Chilled water Delta T (°C).
  Compressor power (amps, voltage)—for DX unit.
  Energy efficiency ratio—for DX unit.

- Fans.
  Fan airflow rate (m³/hr).
  Motor power (amps, voltage).
  VFD data

- Cooling towers.
  Water temperature in/out (°C).
  Fan airflow rate (m³/hr).
  Motor power (amps, voltage).
  VFD data..

- Air cooled chiller.
  Compressor power (amps, voltage)
  Energy efficiency ratio
  Chilled water (pressure drop)
  Chilled water temperature in/out (°C).
  Chilled water Delta T (°C).
  Refrigerant charge—pressure.
  Condenser fan—Motor power (amps, voltage)
  VFD data.

- Heat exchanger.
  Water temperature in/out (°C)—hot and cold sides.
  Water pressure drop (kPa)—hot and cold sides.

- Water cooled chillers.
  Compressor power (amps, voltage)
  Energy efficiency ratio
  Chilled water (pressure drop)

Condenser (pressure drop)
Chilled water temperature in/out (°C).
Chilled water Delta T (°C).
Refrigerant charge—pressure.
VFD data

- Direct expansion air conditioning units.
  Compressor power (amps, voltage)
  Energy efficiency ratio

- Economizers.
  Mixing airflow rate (m³/hr).
  Mixing outlet airflow temperature (°C).

- Air compressors.
  Air pressure (PSI)

2. Water supply system:
   - Pumps.
     Pump water flow rate (lps).
     Motor power (amps, voltage).
     Pump head (kPa)

   - Heat exchangers.
     Water temperature in/out (°C)—hot and cold sides.
     Water pressure drop (kPa)—hot and cold sides.

   - PRV stations.
     On PRV pressure (bar)
     Off PRV pressure (bar)

   - Boilers.
     Boiler efficiency.
     Electric heater power (kW)—for electric powered boilers.
     Fuel consumption efficiency—for fuel powered boilers.
     Water temperature (°C).

3. Drainage system
   - Sump pumps (sewage).

Pump water flow rate (lps).
Motor power (amps, voltage).
Pump head (kPa)
VFD data

4. Storm water management system.
   • Rain water pumps.
   Pump water flow rate (lps).
   Motor power (amps, voltage).
   Pump head (kPa)
   VFD data

5. Building Transportation System
   Elevators.
   Travelators.
   Escalators.
   Motor Power (Amps, Voltage).

6. Fire Fighting Systems
   • Fire Pumps
   Pump water flow rate (lps).
   Motor Power (Amps, Voltage).
   Pump Head (kPa).
   VFD Data

7. Electrical Systems:
   • Motor Control Centers.
   In/Out Voltage at constant frequency (V).
   In/Out Current at constant frequency (Amps).
   Display frequency and actual out voltage
   Ambient temperature
   Operating temperature at constant frequency

   • Variable Frequency Drive (VFD).
   In/Out Voltage at constant frequency (V).
   In/Out Current at constant frequency (Amps).
   Display frequency and actual out voltage
   Ambient temperature
   Operating temperature at constant frequency

8. Lighting.
     Lighting Lux Level.

9. Building Management System
- 2 Way - 3 Way Valve Functionality.
  Control logic.

- Differential Pressure Switch—DPS.
  DPS Set Value.
  Actual pressure measurement.
  Control logic

- Differential Pressure Transmitter—DPS.
  DPT Set Value.
  Actual pressure measurement.
  Control logic

- Airflow/Velocity Meters.
  Air velocity (m³/hr).
  Control logic

- On Coil Temperature & Humidity Sensors.
  Sensor ohmic value and actual temperature value.

- Off Coil Temperature & Humidity Sensors.
  Sensor ohmic value and actual humidity value.

- Space Temperature and Humidity Sensors.
  Sensor ohmic value and actual humidity and
  temperature values.

- Thermostat Functionality.
  Design set point.
  Control logic

SOURCES OF DATA

Data collection and information required for the original equipment design is the baseline for comparing the current operational behavior of the equipment with the intended design function. Therefore, the list of data mentioned in the previous section should be collected from the following sources:

- Drawings: Construction drawings, and equipment specifications.
- Manufacturer O&M Manuals: records of all manufacturers operation and maintenance instructions.
- Testing and Commissioning Records: All testing and commissioning reports including readings about operational parameters of the equipment
- Maintenance Records: Records of maintenance conducted during operation including consumables used (for example, lubricants, fuel, and filters).
- Procedures: all operating or maintenance procedures identified for each equipment.
- Design Information: All design data that indicates the intended function of each equipment and designed operational parameters (for example, expected chilled water flow rate from a pump)
- Original Testing and Commissioning Data: (a complete set of T&C data which will be used as a benchmark to compare the current equipment performance to the original T&C)
- Re-commissioning Data (If available).

When reviewing the above records, refer to the following guidelines:

Validate all data.
Verify that the equipment identification is correct.
Verify that the equipment description is correct.
Verify that the equipment specifications are correct.
Verify that the operational parameters are defined.

Chapter 6

ECM Process—
ECM Inspections

STEP 3. IDENTIFY ECM INSPECTIONS, FREQUENCY,
CRAFT, TOOLS, AND JOB DURATION

Energy centered maintenance model is a proactive model that
focuses on preventive and predictive maintenance types rather than reactive maintenance. The energy-related maintenance inspections should
be combined with the regular preventive maintenance plans of any
equipment. Therefore, when maintenance personnel conduct the regular periodic maintenance for equipment, they will also hold the energy
centered maintenance inspections and do the required measurements,
accordingly if an improvement to the equipment performance it can be
determined.

Successful implementation of a maintenance inspection is critical
to a smooth running energy centered maintenance model within a facility and offers a high number of opportunities to improve equipment
efficiency.

The frequency of conducting energy centered maintenance inspections differ based on the type of equipment and type of maintenance. It
could be sufficient to perform the energy centered maintenance check
annually for one type of equipment and semi-annually for another,
based on equipment type and its energy classification code.

Energy centered maintenance inspections are defined for all
equipment listed in the previous section. The frequency of the inspection should be conducted will also be specified based on the available
reliability maintenance records, available information about equipment
performance and efficiency and based on equipment's energy classification code.

This section represents the core maintenance practice in the energy
centered maintenance model. For specific items of equipment, it is re-

quired to collect the data that are related to the operational behavior of the equipment, and it is, therefore, essential that the maintenance staff should be familiar with the equipment O&M requirements to comply with the maintenance inspections specified therein.

The equipment's energy centered maintenance inspections should be carried out by the appropriately qualified staff who are capable of conducting the inspections and recording the measurements accurately and can make the right judgment about each equipment behavior.

MAINTENANCE RECORDS

Energy centered maintenance inspections and frequency are influenced by the reliability maintenance history of the equipment. The records are necessary to define what maintenance tasks should be conducted as part of preventive and predictive maintenance regimes one the inspection is completed and the current operational performance of the equipment is measured.

A frequently failed asset in maintaining its operational efficiency may require a different energy centered maintenance strategy to be applied than other assets, even if it is performed for the same type (i.e. two motors serving two different air handling units). Or it may be not operationally feasible or is too expensive to maintain. Hence it should be replaced. This type of decision can be judged based on the available maintenance records of each particular asset or equipment within the facility, and based on the cost effectiveness of the remedies of a certain deficiency.

Another important point is the examination of the asset over its lifecycle, the records of previous maintenance contribute in calculating the total cost of ownership of the equipment since installation date.

Maintenance records should include:
— A coherent equipment repair history.
— A record of maintenance performed on equipment.
— The cost of maintenance.
— The cost of energy.
— Replacement information.
— Modification information.
— Spare parts replacement.
— Diagnostic monitoring data (if available in BMS).

— Condition assessment.
— Energy efficiency records.
— Retro-commissioning records.

Maintenance records can be used for activities such as energy efficiency analysis, energy centered maintenance inspections, preventive maintenance tasking, predictive maintenance tasking, frequency planning, and life cycle analysis.

Equipment repair history data are essential to support maintenance activities, upgrade maintenance programs, optimize equipment performance, improve equipment efficiency, plan corrective maintenance, track equipment modifications, and develop equipment and system life-cycle plans.

ENERGY CENTERED MAINTENANCE INSPECTIONS

Identifying energy centered maintenance inspection is the primary process in this guidelines, the inspection should be properly defined, assigned and sequenced for each type of equipment.

Energy centered maintenance checks should not be carried out separately from the regular reliability maintenance but should be combined with the existing preventive and predictive maintenance plans for the equipment.

The aim of this model is to amend the current reliability maintenance job plans to include those inspections that are related to the operational parameter of the equipment (i.e. efficiency). ECM inspections are defined at the end of this section for the following system's equipment:

1. Heating, Ventilation and Air Conditioning System.
2. Water Supply System.
3. Drainage System.
4. Storm Water Management System
5. Building Transportation System
6. Fire Fighting System.
7. Electrical System.
8. Building Management System.

Specifying energy centered maintenance inspections should be established based on clear targets, for example, in an Air Handling Unit the target is to ensure the AHU is capable of delivering the required airflow rate as intended in the design stage, this goal sets what maintenance inspections should be conducted.

Energy centered maintenance inspections help to define what elements of maintenance are required for specific equipment, an outage inspection should be specified for testing and should provide management with information necessary to control equipment performance.

The inspections should be prepared considering the following:

— Determine which deficiencies may have an impact on customer satisfaction to correct.
— Determine which deficiencies are the most cost beneficial to correct.
— Determine which deficiencies are the most performance adequate to correct.
— Determine which performance parameters are critical for equipment operation to correct.

Scheduling energy centered maintenance inspections should be performed in such a way that energy centered maintenance tasks are conducted in the proper sequence, efficiently, and within prescribed time limits.

ENERGY CENTERED MAINTENANCE INSPECTION FREQUENCY

Preventive maintenance is a time-based maintenance that should be performed with pre-determined plans and frequencies, with job plans specifying how often inspection of equipment should take place. Predictive maintenance is a condition based maintenance where any reduction in equipment efficiency (i.e. supplied flow rate) can be captured at an early stage before a deficiency in equipment performance occurs. The frequency of energy centered maintenance inspections could vary based on different parameters such as:

— Equipment Operating Life.
— Physical Condition.
— Failure Interval and Failure Rate.

The more frequent the maintenance inspections take place, the higher the cost but, the greater chances of maintaining equipment efficiency. Conversely the less frequent the inspection, the less the cost but, the higher chances of increased energy use and increased energy waste intervals which result in high corrective maintenance cost. A balance between energy centered maintenance inspection, frequency, cost, and equipment efficiency should be assessed while defining the optimum ECM frequency.

In reliability maintenance, the frequency of performing preventive maintenance tasks can be calculated based on the probability of failure of a machine, or based on failure rate and failure intervals, but this is for those failures which stop the equipment from performing its intended function. Energy centered maintenance is not related to this kind of failures, ECM is focused on the efficiency of the equipment while it is working, therefore calculating specific frequencies to conduct ECM inspections are critical for cost effectiveness.

The energy centered maintenance strategy calls to perform ECM inspections as part of equipment's regular preventive maintenance job plans; it will be either based on the following:

— Annual basis:
Where ECM inspections will be part of the annual PPM plans.

— Semi-Annual basis:
Where ECM inspection will be part of the semi-annual PPM plans.

The operation and maintenance team can revisit the energy centered maintenance inspection frequencies provided within this model depending on the actual condition of the equipment. An increase or decrease in maintenance frequencies should be analyzed based on justifications such as condition monitoring. For example, when there is no loss in equipment efficiency (i.e. pump flow rate), the energy centered maintenance frequency may be reduced.

ENERGY CENTERED MAINTENANCE CRAFT, TOOLS, AND DURATION

Proper planning for the required level of Craft Personnel should be determined as part of the job plan. The productivity of the works done by the maintenance personnel can be optimized if the job plans specify what kind of tools are required while performing the ECM inspections, and what is the time duration needed to complete it.

This section sets the criteria for selecting the right personnel, tools and time duration required to perform energy centered maintenance inspections.

Craft Personnel

Energy centered maintenance inspections and job plans should be performed by a team of appropriately qualified and experienced personnel to achieve safe, and efficient maintenance operations.

The facilities management team should provide the administrative and functional structure that determines the skills needed, appropriate assignments, and the performance standards for each craft group. Maintenance personnel must be skilled inefficient troubleshooting of equipment problems and must be acquainted with plant policies, procedures, systems, and equipment changes that affect their activities. Crafts personnel are usually very knowledgeable about the equipment that they maintain and understand the needs to do a quality job.

Maintenance professionals must consider all of the various requirements for scheduled and unscheduled maintenance; then compile an initial projection of the personnel requirements for the future maintenance activities. Maintenance and operations personnel are intimately familiar with the facility, systems, and equipment. Therefore, they should communicate any scheduling or task deficiencies to supervisors.

Experienced maintenance personnel should meet the following criteria:
— Understand general facility systems and equipment layout.
— Comprehend the purpose and importance of the facility's systems and equipment.
— Understand the effect of ECM work on the facility's systems.
— Assimilate industrial safety, including hazards associated with specific systems and equipment.
— Understand job-specific work practices.
— Comprehend maintenance policies and procedures.
— Be familiar with the personal protective equipment.
— Be capable to evaluate the performance of the equipment.

Craft Personnel Training

Formal training programs should be implemented to develop and improve the knowledge and skills necessary to perform assigned functions and tasks associated with energy centered maintenance model.

These training programs should be based on identified needs and must include provisions for the systematic evaluation of training effectiveness.

The training required for conducting ECM inspection is related to the testing and commissioning of the equipment; the maintenance professionals should define the training needed for their staff based on system needs.

An appropriate skill level requirement should be assigned to each ECM inspection according to its associated function as shown in Table 6.1.

Tools and Special Equipment

Identifying tools and specialized equipment that are required to execute an efficient energy centered maintenance inspections is an essential process that should be planned during the development of ECM job plans and inspections. A controlled supply of the proper type, quality, and quantity of tools and special equipment serves to avoid delays in maintenance work activities and increase worker efficiency. Defining the right tools is essential to allow the maintenance personnel to measure the current performance of the equipment which identifies if any energy waste is noticed and to determine if the equipment is under performing, over performing or performing as intended.

Effective control on ECM inspections is achieved when the appropriate tools and specialized equipment are available to the maintenance team for the timely and accurate execution of inspection. Policies and procedures must be in place that specifically describes the responsibilities and techniques for receiving, inspecting, handling, storing, retrieving, and issuing tools and equipment.

Tools and special equipment area include all tools, special condition-monitoring equipment, diagnostic and check out equipment, calibration equipment, measuring equipment, and so on. Tools and equipment in good condition that are easily obtainable are essential elements for reaching maintenance productivity and service goals. An appropriate tool requirement should be assigned to each ECM task according to its associated function as shown in the following table.

Tools and special equipment control systems should be periodically evaluated for their effectiveness. Specialized equipment and tools for measuring and testing should be calibrated and controlled. All phases of procuring, receiving, inspecting, handling, storing, retrieving, and ensuring of tools, and equipment should be controlled.

Table 6.1 Craft Personnel—Function Description

Function	Craft Personnel	Function Description
Heating, Ventilation, and Air Conditioning	Mechanical Technician, Supervisor, Engineer.	Responsible for the installation, maintenance, and repair of all building ventilating, heating, refrigerating, and cooling systems.
Plumbing	Mechanical Technician, Supervisor, Engineer.	Responsible for the maintenance, and repair of domestic water, steam, sewer, and other utility systems.
Fire Protection	Mechanical Technician, Supervisor, Engineer.	Responsible for the inspection, testing, maintenance, and repair of installed fire suppression systems (fire pumps)
Electrical	Electrical Technician, Supervisor, and Engineer	Responsible for the installation, maintenance, or repair of equipment for the generation, distribution, control, or utilization of electric energy.
Building Management System	Low voltage and control Technician, Supervisor, Engineer.	Responsible for installation and maintenance of all building management system components, communication, and control logic.

Table 6.2. Tools and Special Equipment—Sample List

System	Example Tool	Tool Description
HVAC Plumbing Fire Fighting	Thermometer	A device that measures a temperature gradient in space or a temperature of a liquid, gas or air. Measured in °C or equivalent.
	Anemometer	A device that measures an air flow rate and air speed inside ducts and air terminals. In m³/hr or equivalent.
	Manometer	A device that is used to measure the pressure of a liquid or gas and indicates the difference between two pressures. Measured in PSI or equivalent.
	Combustion Analyser	A device that is used to measure different parameters related to combustion such as combustion air temperature, fuel to air ratio, combustion efficiency, etc.
	Ultrasonic Leak Detector	A device that is used to identify any leaks in a compressed air system such as boilers, or a compressed refrigerant system such as in refrigeration systems.
	Flow meter	A device used to measure the flow rate of a liquid such as a pump flow rate. Measured in m³/hr or equivalent.
Electrical	Multi-meter	A device used to measure electric current, voltage, and resistance over several ranges of value. Measured in Watts, Ohms, or equivalent.
	Power Analyser	A device that measures electrical power characteristics of electrical equipment, it provides precise measurements of real power, power factor, harmonics, and efficiency.
	Electricity Meter	A device that is used to gauge the amount of electricity consumed by a particular electrical equipment such as motors, or lighting. Measured in kWh
	Meggar Tester	A device that is used to gauge the insulation resistance of an electrical cable and measured in kiloohms, megaohms, or gigaohms.
	Voltmeter	A device that is used to gauge the voltage in an electrical circuit or equipment. Measured in Volts
	Thermal Camera	A device using infrared radiation to identify hot spots within electrical panels and electrical cables.
BMS	Ohmmeter	An electronic device to measure resistance in an electronic component or circuit.
	DDC Simulator	A device used to test the functionality of equipment's operation.
	Logic Simulator	Tools used to check the particular function of a logic related to a certain equipment or system

CALIBRATION PROGRAM

The maintenance team should have a documented program for the control and calibration of test equipment and tools that ensure the availability of calibrated specialized equipment and tools.

The calibration program must assure:

—   All calibration standards used by the calibrating agency are traceable to national or recognized standards.

—   All measuring and testing equipment are kept in especially dedicated facilities to control storage, calibration, and issuance.

—   Any critical equipment that was calibrated with out-of-tolerance test equipment is evaluated on time and re-calibrated as necessary.

—   Any tool or measuring and test equipment with actual or suspected defects is marked and isolated to prevent its use.

—   Calibration frequencies help maintain measuring and testing equipment accuracy and availability.

—   The procedures used to calibrate measuring, and testing equipment and tools include records for accountability and traceability of use.

INSPECTION DURATION

Inspection duration specifies the expected time in conducting energy centered maintenance inspections for each equipment (i.e. 2 hrs to do ECM inspection on AHU). Time management data are necessary to quantify operational process shortfalls, calculating the cost of maintenance, and return on investment.

The main objective in specifying the timeline for completing the ECM inspection is the following:

—   Optimizing planning and scheduling functions:
    Setting inspection duration provides the basis for better scheduling and planning of energy centered maintenance plans.

— Measure Service Provider Productivity:
By specifying the expected duration of each inspection, will allow the FM team to identify the productivity of service providers by comparing the planned duration to the actual one. It will enable us to measure wrench time, non-productive time or overtime that the technician spent doing the work.

— Identify actual cost of maintenance:
Specifying the duration of inspections provides the basis for calculating the manpower cost while conducting the ECM inspections; cost measurement is essential for return on investment calculations and cost effectiveness that is assigned with ECM model.

Notes:

Duration of energy centered maintenance inspections might differ from one site to another based on maintainability and accessibility of the equipment. Therefore, the duration that is specified in this guidelines is indicative. However, it should be specified in sites maintenance plans.

The duration of inspection should count for the maintenance personnel productivity. (i.e. time spent to reach the equipment is not counted in wrench time)

ENERGY CENTERED MAINTENANCE INSPECTION PLANS

This section lists down all energy centered maintenance tasks and job plans for all equipment listed in the previous chapter.

Heating, Ventilation and Air Conditioning System

Table 6.3 Energy Centered Maintenance Inspection Plans

Table 6.3.1 Air Handling Units

Equipment Type: Air Handling Units				
Preventive Maintenance				
Inspection	Frequency	Tool	Craft	Duration
Measure airflow rate (m³/hr)	Annual	Anemometer	Mechanical Technician	20-30 min
Check motor's full load current (Amps)	Quarterly	Multi-meter	Electrician	10-20 min
Measure Cooling Coil Performance on Full Load (Air Off Coil Temperature °C)	Annual	Thermometer	Mechanical Technician	10-15 min
Measure Cooling Coil Performance (Pressure Drop, PSI)	Annual	Manometer	Mechanical Technician	10-15 min
Measure Variable Frequency Drive effectiveness	Annual	Multi-meter	Electrician	45-60 min
Measure 2-way/3-way control valves response to space temperature	Quarterly	DDC Simulator	Control Technician/ Electrician	20-30 min
Measure chilled water temperature difference (Delta-T) °C	Quarterly	Thermometer	Mechanical Technician	15-20 min
Predictive Maintenance				
Inspection	Frequency	Tool	Craft	Duration
Measure airflow rate (m³/hr)	Continues	Flow-meter (Connected on BMS)	BMS Operator	5-10 min
Measure air off coil temperature (°C)	Continues	Thermometer (Connected on BMS)	BMS Operator	5-10 min
Measure Fan's Motor Power Consumption (kWh)	Continues	Electricity Meter	BMS Operator	5-10 min
Measure chilled water temperature difference (Delta-T) °C	Continues	Thermometer	Mechanical Technician	5-10 min

Table 6.3.2 Fan Coil Units

Equipment Type: Fan Coil Unit				
Preventive Maintenance				
Inspection	Frequency	Tool	Craft	Duration
Measure airflow rate (m³/hr)	Annual	Anemometer	Mechanical Technician	20-30 min
Check motor's full load current (Amps)	Quarterly	Multi-meter	Electrician	10-20 min
Measure Cooling Coil Performance on Full Load (Air Off Coil Temperature °C)	Annual	Thermometer	Mechanical Technician	10-15 min
Measure Cooling Coil Performance (Pressure Drop, PSI)	Annual	Manometer	Mechanical Technician	10-15 min
Measure Variable Frequency Drive effectiveness	Annual	Multi-meter	Electrician	45-60 min
Measure 2-way/3-way control valves response to space temperature	Quarterly	DDC Simulator	Control Technician/Electrician	20-30 min
Measure chilled water temperature difference (Delta-T) °C	Quarterly	Thermometer	Mechanical Technician	15-20 min
Predictive Maintenance				
Inspection	Frequency	Tool	Craft	Duration
Measure airflow rate (m³/hr)	Continues	Flow-meter (Connected on BMS)	BMS Operator	5-10 min
Measure air off coil temperature (°C)	Continues	Thermometer (Connected on BMS)	BMS Operator	5-10 min
Measure Fan's Motor Power Consumption (kWh)	Continues	Electricity Meter	BMS Operator	5-10 min
Measure chilled water temperature difference (Delta-T) °C	Continues	Thermometer	Mechanical Technician	5-10 min

Note: Predictive maintenance measures for Fan coil units may be implemented for high energy consumption FCUs. It may not be practical to implement it on all FCUs within the facility.

Table 6.3.3 Energy Recovery Units (i.e. Heat Wheels)

Equipment Type: Energy Recovery Units				
Preventive Maintenance				
Inspection	Frequency	Tool	Craft	Duration
Measure airflow rate (m³/hr)	Annual	Anemometer	Mechanical Technician	20-30 min
Check motor's full load current (Amps)	Quarterly	Multi-meter	Electrician	10-20 min
Measure Energy Recovery Performance on Full Load (Air Off Coil Temperature °C)	Annual	Thermometer	Mechanical Technician	10-15 min
Measure Cooling Coil Performance (Pressure Drop, PSI)	Annual	Manometer	Mechanical Technician	10-15 min
Predictive Maintenance				
Inspection	Frequency	Tool	Craft	Duration
Measure airflow rate (m³/hr)	Continues	Flow-meter (Connected on BMS)	BMS Operator	5-10 min
Measure air off coil temperature (°C)	Continues	Thermometer (Connected on BMS)	BMS Operator	5-10 min
Measure Fan's Motor Power Consumption (kWh)	Continues	Electricity Meter	BMS Operator	5-10 min

Table 6.3.4 Boilers

Equipment Type: Boilers				
Preventive Maintenance				
Inspection	Frequency	Tool	Craft	Duration
Measure fuel combustion efficiency	Bi-annual	Combustion Analyser	Mechanical Technician	30-45 min
Inspect steam leakage	Monthly	Ultrasonic Steam Leak Detector	Mechanical Technician	30-45 min
Predictive Maintenance				
Inspection	Frequency	Tool	Craft	Duration
Outlet Water Temperature (°C)	Continues	Thermometer (Connected on BMS)	BMS Operator	5-10 min
Primary System Water Pressure (PSI)	Continues	Manometer (Connected on BMS)	BMS Operator	5-10 min

Table 6.3.5 Pumps

Equipment Type: Chilled Water Pumps				
Preventive Maintenance				
Inspection	Frequency	Tool	Craft	Duration
Measure full speed water flowrate (gpm, lps)	Annual	Anemometer	Mechanical Technician	60-90 min
Check motor's full load current (Amps)	Quarterly	Multi-meter	Electrician	10-20 min
Measure Variable Frequency Drive effectiveness	Annual	Multi-meter	Electrician	45-60 min
Predictive Maintenance				
Inspection	Frequency	Tool	Craft	Duration
Measure water flowrate (gpm, lps)	Continues	Flow-meter (Connected on BMS)	BMS Operator	5-10 min
Measure Pump's Motor Power Consumption (kWh)	Continues	Electricity Meter (Connected on BMS)	BMS Operator	5-10 min
Motor running current (Amps)	Continues	Multi-meter	BMS Operator	10-15 min

Table 6.3.6 Close Control Units

Equipment Type: Close Control Units				
Preventive Maintenance				
Inspection	Frequency	Tool	Craft	Duration
Measure airflow rate (m³/hr)	Annual	Anemometer	Mechanical Technician	20-30 min
Check fan motor's full load current (Amps)	Quarterly	Multi-meter	Electrician	10-20 min
Check DX unit compressor full load current (Amps)	Quarterly	Multi-meter	Electrician	10-20 min
Measure Cooling Coil Performance on Full Load (Air Off Coil Temperature °C)	Annual	Thermometer	Mechanical Technician	10-15 min
Measure Cooling Coil Performance (Pressure Drop, PSI)	Annual	Manometer	Mechanical Technician	10-15 min
Measure Variable Frequency Drive effectiveness	Annual	Multi-meter	Electrician	45-60 min
Measure 2-way/3-way control valves response to space temperature	Quarterly	DDC Simulator	Control Technician/ Electrician	20-30 min
Measure chilled water temperature difference (Delta-T) °C	Quarterly	Thermometer	Mechanical Technician	15-20 min
Measure DX unit Energy Efficiency Ratio EER	Quarterly	Multi-meter	Mechanical Technician/Electrician	30-45 min
Predictive Maintenance				
Inspection	Frequency	Tool	Craft	Duration
Measure airflow rate (m³/hr)	Continues	Flow-meter (Connected on BMS)	BMS Operator	5-10 min
Measure air off coil temperature (°C)	Continues	Thermometer (Connected on BMS)	BMS Operator	5-10 min
Check DX unit compressor full load current (Amps)	Continues	Electricity Meter	BMS Operator	5-10 min
Measure chilled water temperature difference (Delta-T) °C	Continues	Thermometer	Mechanical Technician	5-10 min

Table 6.3.7 Fans

Equipment Type: Fans				
Preventive Maintenance				
Inspection	Frequency	Tool	Craft	Duration
Measure airflow rate (m³/hr)	Annual	Anemometer	Mechanical Technician	20-30 min
Check motor's full load current (Amps)	Quarterly	Multi-meter	Electrician	10-20 min
Measure Variable Frequency Drive effectiveness	Annual	Multi-meter	Electrician	45-60 min
Predictive Maintenance				
Inspection	Frequency	Tool	Craft	Duration
Measure airflow rate (m³/hr)	Continues	Flow-meter (Connected on BMS)	BMS Operator	5-10 min
Measure Pump's Motor Power Consumption (kWh)	Continues	Electricity Meter (Connected on BMS)	BMS Operator	5-10 min

Equipment Type: Cooling Towers				
Preventive Maintenance				
Inspection	Frequency	Tool	Craft	Duration
Check fan motor's full load current (Amps)	Bi-Annual	Multi-meter	Electrician	20-30 min
Cooling Tower Range (In-Out Water Temperature) – Full Load (°C)	Monthly	Thermometer	Mechanical Technician	10-20 min
Measure Variable Frequency Drive effectiveness	Annual	Multi-meter	Electrician	45-60 min
Predictive Maintenance				
Inspection	Frequency	Tool	Craft	Duration
Check fan motor's current (Amps)	Continues	Electricity Meter (Connected on BMS)	BMS Operator	5-10 min
Cooling Tower Range (In-Out Water Temperature) – Full Load (°C)	Continues	Thermometer (Connected on BMS)	BMS Operator	5-10 min

Table 6.3.8 Air Cooled Chillers

Equipment Type: Air Cooled Chillers				
Preventive Maintenance				
Inspection	Frequency	Tool	Craft	Duration
Check Compressor motor's full load current (Amps)	Bi-Annual	Multi-meter	Electrician	10-20 min
Check Condenser Fan motor's full load current (Amps)	Annual	Multi-meter	Electrician	10-20 min
Evaporator pressure drop (PSI)	Bi-Annual	Manometer	Mechanical Technician	30-45 min
Refrigerant leaks test	Quarterly	Ultrasonic Leak Detector	Mechanical Technician	30-45 min
Measure Variable Frequency Drive effectiveness	Annual	Multi-meter	Electrician	45-60 min
Predictive Maintenance				
Inspection	Frequency	Tool	Craft	Duration
Check Compressor motor's current (Amps)	Continues	Electricity Meter (Connected on BMS)	BMS Operator	5-10 min
Chilled water supply temperature (°C)	Continues	Thermometer (Connected on BMS)	BMS Operator	5-10 min
Operating pressure (PSI)	Continues	Manometer (Connected on BMS)	BMS Operator	5-10 min

Table 6.3.9 Water Cooled Chillers

Equipment Type: Water Cooled Chillers				
Preventive Maintenance				
Inspection	Frequency	Tool	Craft	Duration
Check Compressor motor's full load current (Amps)	Bi-Annual	Multi-meter	Electrician	10-20 min
Evaporator pressure drop (PSI)	Bi-Annual	Manometer	Mechanical Technician	30-45 min
Condenser pressure drop (PSI)	Bi-Annual	Manometer	Mechanical Technician	30-45 min
Refrigerant leaks test	Quarterly	Ultrasonic Leak Detector	Mechanical Technician	30-45 min
Measure Variable Frequency Drive effectiveness	Annual	Multi-meter	Electrician	45-60 min
Predictive Maintenance				
Inspection	Frequency	Tool	Craft	Duration
Check Compressor motor's current (Amps)	Continues	Electricity Meter (Connected on BMS)	BMS Operator	5-10 min
Chilled water supply temperature (°C)	Continues	Thermometer (Connected on BMS)	BMS Operator	5-10 min
Operating pressure (PSI)	Continues	Manometer (Connected on BMS)	BMS Operator	5-10 min

Table 6.3.10 Heat Exchangers

Equipment Type: Heat Exchangers				
Preventive Maintenance				
Inspection	Frequency	Tool	Craft	Duration
Measure pressure drop (PSI)	Bi-Annual	Manometer	Mechanical Technician	60-90 min
Heat exchanger effectiveness (%) - (In/Out Temperature)	Bi-Annual	Thermometer	Mechanical Technician	20-30 min
Predictive Maintenance				
Inspection	Frequency	Tool	Craft	Duration
Chilled water supply temperature (°C)	Continues	Thermometer (Connected on BMS)	BMS Operator	5-10 min

Table 6.3.11 Direct Expansion Air Conditioners (DX Units)

Equipment Type: DX Units				
Preventive Maintenance				
Inspection	Frequency	Tool	Craft	Duration
Compressor motor's full load current (Amps)	Bi-Annual	Multi-meter	Electrician	10-20 min
Refrigerant leaks test	Quarterly	Ultrasonic Leak Detector	Mechanical Technician	30-45 min
Predictive Maintenance				
Inspection	Frequency	Tool	Craft	Duration
Check Compressor motor's current (Amps)	Continues	Electricity Meter (Connected on BMS)	BMS Operator	5-10 min

Economizers

Table 6.3.12 Economizers

Equipment Type: Economizers

Preventive Maintenance

Inspection	Frequency	Tool	Craft	Duration
Supply air flow rate after mixing (m³/hr)	Bi-Annual	Anemometer	Mechanical Technician	20-30 min
Supply air flow rate temperature after mixing (°C)	Bi-Annual	Thermometer	Mechanical Technician	20-30 min

Predictive Maintenance

Inspection	Frequency	Tool	Craft	Duration
Supply air flowrate after mixing (m³/hr)	Continues	Anemometer (Connected on BMS)	BMS Operator	5-10 min
Supply air flow rate temperature after mixing (°C)	Continues	Thermometer (Connected on BMS)	BMS Operator	5-10 min

Air Compressors

Table 6.3.13 Air Compressors

Equipment Type: Air Compressors				
Preventive Maintenance				
Inspection	Frequency	Tool	Craft	Duration
Produced air pressure (PSI)	Bi-Annual	Manometer	Mechanical Technician	20-30 min

Water Supply System

Table 6.3.14
Domestic Water Pump Set, Irrigation Pump, and Water Features Pumps

Equipment Type: Pumps				
Preventive Maintenance				
Inspection	Frequency	Tool	Craft	Duration
Measure full speed water flowrate (gpm, lps)	Annual	Anemometer	Mechanical Technician	60-90 min
Check motor's full load current (Amps)	Quarterly	Multi-meter	Electrician	10-20 min
Measure Variable Frequency Drive effectiveness	Annual	Multi-meter	Electrician	45-60 min
Predictive Maintenance				
Inspection	Frequency	Tool	Craft	Duration
Measure water flowrate (gpm, lps)	Continues	Flow-meter (Connected on BMS)	BMS Operator	5-10 min
Measure Pump's Motor Power Consumption (kWh)	Continues	Electricity Meter (Connected on BMS)	BMS Operator	5-10 min
Motor running current	Continues	Multi-meter (Connected on BMS)	BMS Operator	5-10 min

Table 6.3.15 Heat Exchangers

Equipment Type: Heat Exchangers				
Preventive Maintenance				
Inspection	Frequency	Tool	Craft	Duration
Measure pressure drop (PSI)	Bi-Annual	Manometer	Mechanical Technician	60-90 min
Heat exchanger effectiveness (%) - (In/ Out Temperature)	Bi-Annual	Thermometer	Mechanical Technician	20-30 min
Predictive Maintenance				
Inspection	Frequency	Tool	Craft	Duration
Chilled water supply temperature (°C)	Continues	Thermometer (BMS)	BMS Operator	5-10 min

Table 6.3.16 Pressure Reducing Valve Stations

Equipment Type: PRV Stations				
Preventive Maintenance				
Inspection	Frequency	Tool	Craft	Duration
Measure On/ Off Pressure on critical PRVs	Annual	Manometer & Calibration Kit	Mechanical Technician	30-45 min per PRV station
Note: only main PRVs should be recalibrated. For example, recalibrate PRVs main on domestic water lines entering floors				

Table 6.3.17 Boilers

Equipment Type: Boilers				
Preventive Maintenance				
Inspection	Frequency	Tool	Craft	Duration
Measure fuel combustion efficiency	Bi-annual	Combustion Analyser	Mechanical Technician	30-45 min
Inspect steam leakage	Monthly	Ultrasonic Steam Leak Detector	Mechanical Technician	30-45 min
Predictive Maintenance				
Inspection	Frequency	Tool	Craft	Duration
Outlet Water Temperature (°C)	Continues	Thermometer (Connected on BMS)	BMS Operator	5-10 min
Primary System Water Pressure (PSI)	Continues	Manometer (Connected on BMS)	BMS Operator	5-10 min

Drainage System

Table 6.3.18 Sump Pumps

Equipment Type: Pumps

Preventive Maintenance

Inspection	Frequency	Tool	Craft	Duration
Measure full speed water flowrate (gpm, lps)	Annual	Anemometer	Mechanical Technician	60-90 min
Check motor's full load current (Amps)	Quarterly	Multi-meter	Electrician	10-20 min
Measure VFD effectiveness	Annual	Multi-meter	Electrician	45-60 min

Predictive Maintenance

Inspection	Frequency	Tool	Craft	Duration
Measure water flowrate (gpm, lps)	Continues	Flow-meter (on BMS)	BMS Operator	5-10 min
Measure Pump's Motor Power Consumption (kWh)	Continues	Electricity Meter (on BMS)	BMS Operator	5-10 min
Motor running current (Amps)	Continues	Multi-meter	BMS Operator	5-10 min

Storm Water Management System

Table 6.3.19 Rain Water Pumps

Equipment Type: Pumps

Preventive Maintenance				
Inspection	Frequency	Tool	Craft	Duration
Measure full speed water flowrate (gpm, lps)	Annual	Anemometer	Mechanical Technician	60-90 min
Check motor's full load current (Amps)	Quarterly	Multi-meter	Electrician	10-20 min
Measure VFD effectiveness	Annual	Multi-meter	Electrician	45-60 min

Predictive Maintenance				
Inspection	Frequency	Tool	Craft	Duration
Measure water flowrate (gpm, lps)	Continues	Flow-meter (on BMS)	BMS Operator	5-10 min
Measure Pump's Motor Power Consumption (kWh)	Continues	Electricity Meter (on BMS)	BMS Operator	5-10 min
Motor running current (Amps)	Continues	Multi-meter	BMS Operator	5-10 min

Building Transportation System

Table 6.3.20 Travelators & Escalators.

Equipment Type: Travelators & Escalators				
Preventive Maintenance				
Inspection	Frequency	Tool	Craft	Duration
Check motor's full load current (Amps)	Quarterly	Multi-meter	Electrician	20-30 min
Auto Start/ Stop Command	Quarterly	Multi-meter	Electrician	20-30 min
Predictive Maintenance				
Inspection	Frequency	Tool	Craft	Duration
Measure Motor Power Consumption (kWh)	Continues	Electricity Meter (Connected on BMS)	BMS Operator	5-10 min

Table 6.3.21 Elevators

Equipment Type: Elevators				
Preventive Maintenance				
Inspection	Frequency	Tool	Craft	Duration
Check motor's full load current (Amps)	Quarterly	Multi-meter	Electrician	20-30 min
Predictive Maintenance				
Inspection	Frequency	Tool	Craft	Duration
Measure Motor Power Consumption (kWh)	Continues	Electricity Meter (Connected on BMS)	BMS Operator	5-10 min

Fire Fighting System

Table 6.3.22 Fire Fighting Pumps

Equipment Type: Pumps				
Preventive Maintenance				
Inspection	Frequency	Tool	Craft	Duration
Measure full speed water flowrate (gpm, lps)	Annual	Anemometer	Mechanical Technician	60-90 min
Check motor's full load current (Amps)	Quarterly	Multi-meter	Electrician	10-20 min
Measure Variable Frequency Drive effectiveness	Annual	Multi-meter	Electrician	45-60 min
Predictive Maintenance				
Inspection	Frequency	Tool	Craft	Duration
Measure water flowrate (gpm, lps)	Continues	Flow-meter (on BMS)	BMS Operator	5-10 min
Measure Pump's Motor Power Consumption (kWh)	Continues	Electricity Meter (on BMS)	BMS Operator	5-10 min
Motor running current (Amps)	Continues	Multi-meter	BMS Operator	5-10 min

Electrical System

Table 6.3.23 Motor Control Center

Equipment Type: MCC				
Preventive Maintenance				
Inspection	Frequency	Tool	Craft	Duration
In/out voltage at constant frequency	Annual	Voltmeter	Electrician	10-15 min
In/out current at constant frequency	Annual	Multi-meter	Electrician	10-15 min
Operating temperature	Quarterly	Thermal camera	Electrician	10-15 min
Predictive Maintenance				
Inspection	Frequency	Tool	Craft	Duration
Verify the out voltage with frequency	Quarterly	Multi-meter	Electrician	10-15 min
Ambient temperature	Continues	Thermometer	Electrician	10-15 min

Table 6.3.24 Variable Frequency Drive

Equipment Type: VFD				
Preventive Maintenance				
Inspection	Frequency	Tool	Craft	Duration
In/out voltage at constant frequency	Annual	Voltmeter	Electrician	10-15 min
In/out current at constant frequency	Annual	Multi-meter	Electrician	10-15 min
Operating temperature	Quarterly	Thermal camera	Electrician	10-15 min
Predictive Maintenance				
Inspection	Frequency	Tool	Craft	Duration
Verify the out voltage with frequency	Quarterly	Multi-meter	Electrician	10-15 min
Ambient temperature	Continues	Thermometer	Electrician	10-15 min

Table 6.3.25 Light Bulbs

Equipment Type: Light Bulbs				
Preventive Maintenance				
Inspection	Frequency	Tool	Craft	Duration
Lux Level (foot-candle)	Annual	Lux Meter	Electrician	10-15 min per light

Building Management System—BMS

Table 6.3.26 Two Way Control Valve

Equipment Type: 2 Way Control Valve				
Preventive Maintenance				
Inspection	Frequency	Tool	Craft	Duration
Check actuator response in line with control signal	Quarterly	Voltmeter/ Visual	BMS and Mechanical Technicians	15-20 min
Check feedback correspondence to control signal	Quarterly	Voltmeter/ Visual	BMS and Mechanical Technicians	15-20 min
Check response time with respect to command	Quarterly	Visual	BMS and Mechanical Technicians	15-20 min
Predictive Maintenance				
Inspection	Frequency	Tool	Craft	Duration
Valves response to AHU/ FCU temperature set point	Continuous	Frontend software	BMS Operator	5-10 min

Table 6.3.27 Differential Pressure Switch

Equipment Type: DPS

Preventive Maintenance				
Inspection	Frequency	Tool	Craft	Duration
Check DPS functionality with respect to control signal	Quarterly	Multi-meter/ Visual	BMS and Mechanical Technicians	15-20 min
Conduct continuity loop testing	Quarterly	Multi-meter	BMS and Mechanical Technicians	15-20 min
Compare differential pressure value on frontend and on site	Quarterly	Calibration Manometer	BMS and Mechanical Technicians	15-20 min

Predictive Maintenance				
Inspection	Frequency	Tool	Craft	Duration
Differential pressure value	Continuous	Frontend software	BMS Operator	5-10 in

Table 6.3.28 Differential Pressure Transmitter

Equipment Type: DPT

Preventive Maintenance

Inspection	Frequency	Tool	Craft	Duration
Check DPT response with respect to control signal	Quarterly	Multi-meter/ Visual	BMS and Mechanical Technicians	15-20 min
Conduct continuity loop testing	Quarterly	Multi-meter	BMS and Mechanical Technicians	15-20 min
Compare differential pressure value on frontend and on site	Quarterly	Calibration Manometer	BMS and Mechanical Technicians	15-20 min

Predictive Maintenance

Inspection	Frequency	Tool	Craft	Duration
Differential pressure value	Continuous	Frontend software	BMS Operator	5-10 Min

Table 6.3.29 Flowrate/Velocity Meters

Equipment Type: Flowrate/ Velocity Meters

Preventive Maintenance

Inspection	Frequency	Tool	Craft	Duration
Compare flowrate value on frontend and actual on-site value	Bi-Annual	Calibration flows meter/ velocity meter	BMS and Mechanical Technicians	15-20 min
Conduct continuity loop testing	Quarterly	Multi-meter	BMS and Mechanical Technicians	15-20 min
Check flowrate response to software command (increase/ decrease)	Quarterly	Multi-meter/ flow meter	BMS and Mechanical Technicians	15-20 min

Predictive Maintenance

Inspection	Frequency	Tool	Craft	Duration
Flowrate/Velocity value	Continuous	Frontend software	BMS Operator	5-10 min

Table 6.3.30 Cooling Coil Temperature and Humidity Sensors

Equipment Type: Cooling Coil Temperature and Humidity Sensors				
Preventive Maintenance				
Inspection	Frequency	Tool	Craft	Duration
Check the sensor' ohmic value and actual temperature where it is installed	Quarterly	Thermometer	BMS Technician	10-15 min
Predictive Maintenance				
Inspection	Frequency	Tool	Craft	Duration
Frontend recorded temperature value	Continuous	Frontend software	BMS Operator	5-10 min

Table 6.3.31 Chilled Water Temperature

Preventive Maintenance				
Inspection	Frequency	Tool	Craft	Duration
Check the sensor' ohmic value and actual temperature where it is installed	Quarterly	Thermometer	BMS Technician	10-15 min
Predictive Maintenance				
Inspection	Frequency	Tool	Craft	Duration
Frontend recorded temperature value	Continuous	Frontend software	BMS Operator	5-10 min

Table 6.3.32 Space/Return Air Temperature and Humidity Sensors

Equipment Type: Space/ Return Air Temperature and Humidity Sensors

Preventive Maintenance

Inspection	Frequency	Tool	Craft	Duration
Check the sensor' ohmic value and actual temperature where it is installed	Quarterly	Thermometer	BMS Technician	10-15 min
Check response of two-way valve to space/ return air temperature	Quarterly	Physical	BMS Technician	10-15 min

Predictive Maintenance

Inspection	Frequency	Tool	Craft	Duration
Frontend recorded temperature value	Continuous	Frontend software	BMS Operator	5-10 min

Table 6.3.33 Control Logic for all equipment controlled by BMS

Equipment Type: Control Logic for all equipment				
Preventive Maintenance				
Inspection	Frequency	Tool	Craft	Duration
Control Logic for all equipments that are part of Energy Centered Maintenance process. For example: AHUs	Quarterly	Logic Simulator	BMS Technician/ Mechanical Technician/ Electrical Technician	20-30 min
FCUs				15-20 min
FAHUs				20-30 min
Heat Recovery Units				15-20 min
Close Control Units				15-20 min
Pumps				60-90 min
Water Features				60-90 min
Chillers				60-90 min
Heat Exchangers				60-90 min
Fans				15-20 min
Lifts and Escalators				15-20 min
Travelators				15-20 min
Boilers				60-90 min
Cooling Towers				60-45 min
DX Units				15-20 min
MCC				15-20 min
VFDs				20-30 min

Chapter 7

ECM Process — Measuring Equipment's Current Performance

STEP 4. MEASURING EQUIPMENT'S CURRENT
PERFORMANCE AND COMPARE TO BASELINE

The maintenance personnel should be capable of measuring the current performance of the equipment during energy centered maintenance inspection. Current performance defines the actual operational condition of the equipment which will be compared to the baseline performance as recorded in the testing and commissioning phase.

Measuring the equipment's current performance involves collecting and analyzing actual data about the operational parameters of the equipment. Parameters such as equipment's efficiency & power consumption should be measured. Those data will then help in determining if any part of the equipment is not delivering its intended function, which results in identifying what kind of corrective actions should be done to improve and restore the operational efficiency of the equipment.

MEASURING EQUIPMENT'S CURRENT PERFORMANCE

Measuring equipment performance is usually performed in conjunction with the planned reliability maintenance job plans (linked with regular PPM inspections), for each energy critical assets (ECC 5, 4 and 3). The requirement is to give an indication of the expected operational condition of each equipment, indicate any possible action requirements to improve or restore the performance of the equipment and enable better planning of future maintenance tasks.

Once the current operational condition of the equipment is conducted for ECC 5, 4 & 3 assets, a comprehensive, balanced, proactive maintenance strategy can be developed to restore and maintain the performance of the asset. This process will:

—   Identify the current operational condition of the assets which will be compared to the baseline data collected about each equipment.
—   Help to identify the cause of performance deterioration and corrective action.
—   Calculate the cost of repair and the cost effectiveness.
—   Improving the level of maintenance.

Measuring current equipment's performance is a purposeful examination of the equipment with the intent to benchmark to baseline information or identify problems. The key to the success lies in the dedication of time for the particular purpose of assessment and the qualifications of the inspector. No intention is made during the operational assessment to repair the identified problems immediately; the workforce merely recognizes them.

The measurements shall be compared with the baseline value, and the maintenance personnel shall be capable of judging if the equipment is over performing, underperforming or performing as intended. For example, measuring motor running current may be acceptable if it is running within an acceptable range compared to testing and commissioning values, or an AHU is delivering an acceptable range of airflow compared to original data.

The principal of regular measurement of equipment efficiency aim of the balanced, proactive maintenance approach is to uncover deficiencies or energy waste before they on early stage to reduce its impact on the overall energy consumption of the facility. The sooner the problems are found, the greater opportunity for planning, gathering materials, coordinating outages, estimating, allocating resources, and restoring equipment performance.

A sample equipment's list, with defined performance measures (i.e. motor consumption) that should be measured, is defined in Chapter 9. Similarly, acceptable performance range is also determined, thus if the equipment is performing within this range, then the equipment is considered as performing as intended. Otherwise, a root-cause analysis should be conducted.

ROOT CAUSE ANALYSIS

The aim of root cause analysis process is to prevent the reoccurrence of a functional failure in reliability maintenance regime. Similarly, in energy centered maintenance the goal is to eliminate operational deficiencies of the equipment and to reduce its energy waste. The cause of the energy waste or functional deficiency must be determined through a thorough investigation. This investigation would include researching equipment history, evaluating current conditions and following RCA methods.

Energy centered maintenance model recommends that all operational deficiencies and recognized energy wastes from all energy critical equipment are investigated to eliminate it or reduce it. Root-cause analysis can assist in minimizing or eliminating energy waste during equipment operation by developing effective corrective actions and by adjusting the balanced proactive maintenance strategy approach to best suit the equipment.

The RCA process is applied uniformly to determine the cause of an operational deficiency or energy waste. This process is established on evidence-based causal thinking. RCA can be done by one person troubleshooting a problem or by a team of people investigating a major event. It is the same process throughout the organization.

There are two types of problems:

— **Rule-Based** problems that have one solution.
— **Event-Based** problems that have many possible solutions.

Most problems that occur in a facility are Event-Based problems, i.e., there are many possible solutions.

RCA is an organized methodology or process for determining the cause and effect of a particular failure (or operational deficiency/energy waste) and finding solutions to minimize or eliminate them from reoccurrence. The RCA process plays and important role in:

— Minimizing or eliminating costly failures or operational deficiencies.
— Developing effective corrective actions.
— Creating a prevention culture based on cause and effect.
— Focusing efforts on effective solutions.

The process of Root Cause Analysis involves the following two steps:

Define the Problem:

The process starts by using the information collected during the inspection which identified a certain problem such as low equipment performance, low energy efficiency, and high energy waste. The problem should then be further investigated by defining the following four main elements:

1.   **What**: the actual problem. This may be the indicator or symptom of the problem. It is possible that there is more than one problem for any given event or incident.

2.   **When**: the actual time the problem occurred and the status of the system or process when the problem occurred.

3.   **Where**: Where the problem occurred. This may be the site, facility, building, system, component, or the equipment ID.

4.   **Significance**: the significance of the problem. The significance relates to the efficiency and energy consumption of the equipment.

Use Cause and Effect Chart

Cause and Effect Chart (so called Fishbone Diagram, or Ishikawa Diagram) is a systematic method for identifying all possible causes for a certain effect or a problem. The diagram helps in suggesting all possible causes about energy wastes or operational deficiencies that are found in equipment during energy centered maintenance inspection. The diagram was created by Karou Ishikawa (1968). Figure 7.1

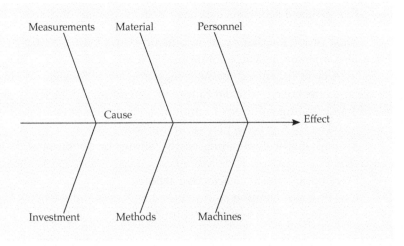

Figure 7.1 Cause and Effect Chart

Once the problem is defined, a cause and effect chart should be prepared by identifying the primary cause of the problem. The causes should then be evaluated by asking "Why" the cause could happen. The process should continue to identify all possible causes of the problem, those causes should then have enough evidence that it is related to the equipment, the causes will then be used to define the required corrective actions and to calculate cost effectiveness to solve the problem.

The motivation for using cause and effect chart is to understand deficiency data and reduce costly energy wastes by appropriate, cost-effective corrective actions. Root-cause analysis and the determination of deficiency and energy waste type, cause, a leads to accurate, efficient, and cost-effective corrective actions.

A sample equipment's list, with defined deficiencies (i.e. motor consumption), are defined in the Chapter 9. The table lists all possible causes against a certain operational problem.

How to use Fishbone Diagram

The fishbone diagram swims to the right. The effect is on the right and can be either a problem or opportunity. For ECM, it is a problem. The top or bottom with personnel, machines, materials, methods, environment and measurements are called the major bones of the fishbone. The major bones' titles are not usually specified. They can come from a process or these lists:

6   Ms-Machines, Methods, Money, Material, Measurement and Management

5   Ps-People, Procedure, Program, Process, Personnel

1   S-System 1E-Environment 3Ts-Tools, Techniques, & Training

The major bones shown in Figure 7.2 are common in determining root causes for problems experienced in ECM. Normally, four major bones are sufficient. Visio has a fishbone diagram and is easy to use. Also, drawing the skeleton (effect and the four major bones) on a white board and then the team or one supervisor or technician can write a possible root cause on a yellow 3M Company's "post it." Then sticking it on the appropriate major bone, and keep doing this until no other possible root causes can be thought of by the participant(s). Next, each possible root cause is evaluated, and the ones that are the strongest possibilities are circled. They are then validated using data where available or a Root Cause Matrix.

Often if insufficient people are available to accomplish all the maintenance and maintenance management duties, someone may be chasing too many rabbits. Running this way and then that way and not getting anything done. To discover potential root causes, they decided to use a root cause analysis, specifically a fishbone diagram to identify why they are having this problem. Figure 7.2 Fishbone Diagram shows this.

Notice all the small bones and how the arrows go into the bones and then the major bones arrows point and go into the effect. The Causes flow in the effect. This is why it is called a cause and effect diagram. There are normally many small bones and only four major bones. In this example, the major bones selected were team/team members, objectives and targets, projects, and measures. Each major bone has several small bones.

The energy wastes that are captured during the implementation of the ECM model can result. Later in this book, an example of energy wastes might have multiple causes. Chapter 9 lists a detailed list of each problem, cause, and effect and suggests ways to eliminate those causes. Energy waste causes during equipment operation can be one of the following types.

Note that this list is not comprehensive:
1. Running when it should be turned off
2. Having to work harder
3. Wear
4. Corrosion
5. Leaks
6. Stuck in the open or closed position
7. Not performing
8. Over performing
9. Overheating
10. Waste build up
11. Incorrect settings
12. Switch or sensor failure (broke or seized)

In the future as organizations implement, ECM this list should grow. In applying ECM later, examples will be shown for each of these energy wastes or excessive uses of energy. ECM includes routine inspections and then fixing any problems encountered.

Figure 7.2 Fishbone Diagram

Chapter 8

ECM Process—Identifying Corrective/Preventive Action & Cost Effectiveness

STEP 5. IDENTIFYING CORRECTIVE/
PREVENTIVE ACTION & COST EFFECTIVENESS

When the root-cause of a certain equipment performance deficiency is known, the right corrective action and repair can be implemented. Energy centered maintenance model focus on identifying any operational deficiencies or energy waste, therefore the corrective action may include specific repairs or replacements, therefore the cost effectiveness of the repair need to be determined prior conducting the corrective action.

IDENTIFYING CORRECTIVE/PREVENTIVE ACTION

Improving equipment efficiency requires proactive involvement in finding the root cause of a deficiency or energy waste and then instituting corrective action for improvements. Corrective/preventive action is an action taken to repair, improve or restore an equipment failure or performance and to prevent it from reoccurrence. The action is decided based on the findings of the root cause analysis of the problem and the associated cause and effect results which aim to prevent the problem from to appear again.

When the maintenance personnel conducts ECM inspections, and they find that certain equipment is underperforming or wasting energy, an immediate corrective action may be undertaken, or it may be scheduled. If the deficiency is critical for the facility operation or equipment's energy waste is high, corrective action should be planned and schedule within 0-48 hours from the time of reporting the problem.

The corrective actions may then be amended in the normal maintenance program by adding new maintenance tasks that provide a proactive approach to prevent the problem from reoccurrence, in this case, the corrective action will be conducted in a proactive manner which makes it a preventive action. Adjustments can be made to the balanced proactive maintenance strategy approach to best suit the equipment.

Energy centered maintenance is a balanced integration of preventive, predictive, and proactive maintenance strategies that emphasizes the elimination of energy wastes, thereby controlling energy consumption and equipment efficiency. The integration between those maintenance strategies with root-cause analysis and corrective/preventive actions leads to improving the maintenance regime and eliminating the problem of reoccurrence.

An example of a corrective action in ECM model is a fan that is delivering lower airflow rate than originally commissioned, the cause maybe a worn bearing and root-cause is the low level of lubrication. In this example, the corrective action is to replace the bearing, while the preventive action is to adjust the normal maintenance program to inspect the bearings and the level of lubrication as part of the planned maintenance.

Identifying corrective action will help in improving the overall balanced, proactive maintenance strategy in the following:
- Identifying what need to be done to restore equipment performance and what are the expected results.
- Identifying preventive actions to prevent the problem from being happening again.
- Identifying which maintenance process needs improvement.
- Identifying if new processes need to implement.
- Identifying new training for maintenance personnel.
- Identifying all costs associated with implementation and determining cost effectiveness.

IDENTIFYING COST EFFECTIVENESS

Maintenance engineers and energy managers are responsible for verifying that appropriate and effective corrective actions have been taken in a cost effective manner.

Before implementing the corrective and preventive actions, it is required to consider the cost effectiveness of the maintenance actions. Some actions may require no or low cost for implementation (i.e. replacing temperature sensor), while other actions may require an investment cost (i.e. replacing motor).

The cost effectiveness of the corrective and preventive actions are calculated based on the cost of implementation compared to the energy cost saved. If the energy cost saved is higher that the corrective and preventive actions cost, the energy centered maintenance implementation is cost effective.

Cost effectiveness should be calculated for each ECM tasks as well as corrective and preventive actions, calculating cost effectiveness should count for all maintenance activities costs associated with those actions, as well as potential energy reduction as defined in a certain period.

The maintenance costs include multiple elements such as:

- Time of maintenance
- Labor costs
- Materials and consumables cost
- Equipment cost
- Calibration cost
- Spare parts cost

Energy conservation is a basic unit of ECM. It is cost effective in that it does not include high labor costs and materials. Most conservation actions are relatively inexpensive or would already be purchased if an energy conservation program were implemented.

Energy saved could be calculated by different ways for each type of equipment by itself. For example, energy saved by enhancing motor efficiency equals the difference in motor's kWh before and after enhancement. This difference is then can be converted to cost saving in a defined period (can be months, years, or life cycle of the equipment) and can be compared to maintenance cost.

The main aim of energy centered maintenance model is to assist the organization in its overall energy program to reduce its energy use and to ensure efficient cost effective operation. If the cost of ECM maintenance actions is less than the impact of the energy waste of the equipment, the appropriate corrective, and preventive maintenance activity should be implemented.

The following formula should assist the maintenance personnel to identify the cost effectiveness of the corrective/preventative actions that need to be implemented to restore the equipment efficiency (i.e. motor power consumption).

Cost effectiveness % = Cost of periodic energy saved –
Cost of maintenance tasks
Cost of periodic energy saved 100%

If the cost-effectiveness value is equal or more than zero, then it is cost effective to conduct the energy centered maintenance tasks and the associated corrective and preventive actions.

On the other hand, if the value is in minus (less than zero), then the cost of maintenance is higher than the cost of energy saved. In this case, the maintenance personnel should account for other factors in determining whether to conduct the ECM tasks or not.

Factors like the contribution of this equipment energy saving to the total building energy saving, impact of the low-performance cost of business operation, the impact of low-performance cost on occupant's comfort level, will be required to evaluate the effectiveness and the need of returning the equipment to its optimum operational efficiency.

RESTORING EQUIPMENT EFFICIENCY

Restoring equipment efficiency aims to operate the equipment in a satisfactory manner. Satisfactory performance implies that specific criteria must be established to describe what is considered as satisfactory operation. Satisfactory operation is achieved when the equipment is operating as per its design intent and as original commissioned, or in more efficient manner.

Corrective actions are required to restore the performance of a non-performing equipment; preventive actions prevents the problem to happen again. It is necessary to verify that the corrective action was effective, not only in eliminating the cause of the deficiency, but also in restoring the equipment's performance. Whatever corrective action is taken, ensure that the original problem is fixed and that no new deficiency is introduced.

Energy centered maintenance focuses on an environment of actions in which equipment efficiency is the main concern. Once ECM inspection is done, and the problem is defined, corrective actions should be implemented to restore the performance of the equipment. Restoring equipment performance is focusing on the following outputs:

- Maximize equipment's operational efficiency.
- Restoring equipment's energy efficiency.
- Reducing energy waste by the equipment.
- Lowers operation cost by reducing energy consumption.
- Restoring the original performance of the equipment.
- Improves equipment quality.
- Ensuring equipment is delivering its intended performance.
- Understanding the effect of equipment age, operating environment on its performance.
- Obtaining data for continuous improvement and high operational effectiveness.
- Corrective/preventive actions are required to restore the performance of a failing equipment. It is necessary to verify that the corrective action was effective, not only in eliminating the cause of the failure, but also in restoring the equipment's performance. Whatever correction/preventive actions are taken, ensure that the original problem is fixed and that no new problems are introduced.

The root-cause analysis process and identification of corrective actions decide what actions are needed to restore the efficiency of the equipment and its performance. Cost effectiveness decides which measures are most cost-effective to be implemented. Restoring equipment efficiency and performance may require a full design review that defines how the performance will be restored. For example, replacing fan motor, or rewinding it.

Implementation of energy centered maintenance tasks in a proactive manner allows the equipment to perform in an efficient manner for a given time when used under specified operating conditions in a given environment. Implementation of ECM tasks results in restoring equipment performance to its original condition and so that the equipment can continue to be used for its intended function.

Chapter 9

ECM Process—Updating Preventative Maintenance Plans

STEP 6. UPDATING PM PLANS ON CMMS

Once the energy centered maintenance approach has been established, regular use of the maintenance program should be implemented in terms of planned preventive maintenance as well as predictive maintenance. The new program should be updated on the balanced proactive maintenance strategy to establish a proactive system that reviews the energy efficiency data of the equipment which helps to identify system operational performance before a significant deficiency occurs.

The new maintenance program should be updated in the maintenance management system (CMMS) to ensure all energy centered maintenance tasks that are defined during ECM inspection are now part of the preventive maintenance plans and predictive maintenance practice.

WHAT IS CMMS?

CMMS is sometimes called CMMIS with the I being "Information" which describes what it is. It is a computerized maintenance management system that contains an organization's maintenance activities. The IS stands for the information system. It is also referred to as the organization's facility maintenance program.

A CMMS software package maintains a computer database of information about an organization's maintenance program and operations. This information is intended to help improve maintenance workers' efficiency and productivity. For example, determining which equipment requires maintenance—what and when and which supply rooms contain the parts or materials needed for a particular work order. And to help management make better decisions (for example, comparing the cost of equipment failure versus preventive maintenance for each machine). CMMS data may also be used to show regulatory compliance and compliance with an ISO requirement.

The following identifies what the CMMS should be able to:

- Address all resources involved in maintenance.
- Maintain maintenance inventory and storeroom location.
- Record and maintain work history of all types of maintenance PM, predictive, emergency, and corrective.
- Include work tasks and frequencies for each craft.
- Effectively interface and communicate with related and supporting systems ranging from work generation through work performance, evaluation and performance reporting.
- Provide feedback information for analysis and decision-making.
- Reduce costs through effective maintenance planning and execution

A modern CMMS meets all these requirements and assists the facilities maintenance manager with, planning, scheduling, control, performance, evaluation, and reporting. CMMS will also maintain historical data for management use and provide meaningful maintenance metrics. Therefore, CMMS provides a work order system, asset management, inventory/purchasing, PM management, work scheduling, and management reports. CMMS is common in the manufacturing industry, facilities both government and civilian, fleet, service providers, oil and gas and other industries. CMMS software packages can be either web-based (hosted by the company selling the product on an outside server), or LAN-based, (The organization buying the software hosts the product on their server.)

CMMS improves mechanic's or a technician's wrench time, enhances spare parts inventory and streamlines procurement of parts and materials. CMMS benefits are:

- Easy to use
- Quick to implement maintenance program
- Minimizes downtime and improves productivity
- Provides maintenance records and history that serves as compliance proof.

CMMS provides preventative maintenance scheduling, work orders, work or service requests, inventory control, predictive maintenance, maintenance reports, and other functions. The CMMS user interface allows for a quick setup, easy data conversion, and the vendor provides training for the users.

CMMS vendors claim CMMS reduces costs and asset downtime

and increases productivity in less than a month. They claim it will extend the life cycle for your facility, decrease your liability, and lower your operating costs without any significant upfront investment.

Provides online planned maintenance scheduling that helps generate, schedule, and manage recurring tasks which are the heart of ECM. Some systems allow sending work order information to maintenance crews in the field, enabling them to receive and complete tasks away from their shop. CMMS is excellent at scheduling jobs, assigning personnel, earmarking materials, recording maintenance costs, and tracking information such as the cause of the problem (if any), downtime that occurred (if any), and suggestions for future action. The CMMS schedules preventive maintenance based on maintenance plans. Different CMMS software packages use different techniques for highlighting when a PM job or task should be performed.

CMMS keeps track of preventive maintenance jobs, including step-by-step instructions. This action is critical for ECM to be successful. CMMS provides an online work order management system that streamlines your work order process, including work request generation, progress and completion status tracking and reports, and reporting of essential maintenance management data and information.

UPDATING PM PLANS ON CMMS

Updating CMMS and maintenance program with specific energy centered maintenance tasks provide vital information to the maintenance and operations personnel that enables them to maintain efficiency and mitigate performance degradation, as well as minimize or eliminate energy wastes associated with the equipment operation.

Energy centered maintenance plans should be created for each energy critical equipment, and it should be integrated with regular reliability maintenance plans. The PM plans should be effective, efficient, and safe to perform based on the strategy (predictive and/or preventive) applied to the equipment. PM plans are also defined as job plans; job plans must be written for every equipment based on energy criticality.

Updating PM plans requires the following:
• Identifying ECM tasks and frequency.
• Identifying and match the appropriate skill sets to the tasks.
• Identifying the appropriate materials to the tasks.

- Identifying the appropriate tools/special equipment to the tasks.
- Identifying all other resources needed to perform the job.
- Upload completed job plans to current CMMS.

The job plans provide all the details regarding safety, environmental, and regulatory issues, as well as the operations, required downtime, affected components/systems, materials, labor, and tools required to do the work. The procedural part of the plan contains a task or a logical sequence of tasks, while each task consists of some related of steps.

The fundamental objective is to conduct efficient, safe and standardized energy centered maintenance job plans as part of regular reliability maintenance plan, which ensures proper, quality PM procedures that maintain the on-going efficiency of equipment.

Integrating ECM job plans with reliability job plans will enhance the productivity of the service provider who is conducting the tasks onsite. Job plans should also be scheduled and tracked to identify the effectiveness of the service provider. Therefore, job plans should be updated in the company's CMMS system.

The use of CMMS will help to develop a site-specific energy centered maintenance program based on equipment functionality, designed to enhance equipment efficiency, make the best use of maintenance resources, and provide correct data for additions or revisions to the existing or new maintenance programs.

PLANNING AND SCHEDULING NEXT INSPECTION

When energy centered maintenance job plans are developed, the FM team shall plan and schedule the next inspection to ensure a proactive approach is taken to maintenance equipment efficiency.

Maintenance planning and scheduling is a significant improvement strategy for maintenance operations and is the single most effective procedure to increase craft labor productivity, effectiveness, and quality. Planning and scheduling work in advance provides the facility the ability to control maintenance activities, reduce costs, reduce the risk of equipment performance deficiency, and improve productivity.

Planning and scheduling emphasize the need to plan and schedule energy centered maintenance activities to maximize the wrench time of the craftspeople doing the work. To create a situation whereby the people who do the tasks arrive at a specified work site fully prepared, with all the necessary instructions, permits, clearances, materials, tools, and equipment to undertake the work order tasks associated with a particular job.

This is a significant improvement strategy for maintenance operations because the work is planned and the right resources are coordinated and are available to do the job correctly the first time. The difference between planning and scheduling is:

Effective planning takes into consideration all of the factors involved in doing the job, along with the sequence in which the factors come into play. By coordinating available resources, effective planning facilitates establishing minimum time and optimum cost work methods.

Efficient scheduling assures a balanced flow of work to the shops by maintaining a proper balance between work capacity and workload.

When the join plans are planned and scheduled, the ECM inspections will take place on a regular basis as part of reliability maintenance plans. Executing maintenance job plans will maintain the efficiency of the equipment and will minimize or eliminate energy waste.

Each site should analyze the potential benefits to be derived from planning and scheduling energy centered maintenance inspection and should develop the plan to implement a planning and scheduling function. This will ensure that:

- ECM inspection is identified and prioritized.
- ECM inspection is accurately planned.
- ECM inspection is scheduled at the right time.
- ECM inspection is assigned and executed in a timely manner.

Energy centered maintenance inspections will help the FM personnel to identify any potential improvements that can be conducted to a certain equipment to ensure it is functioning as originality designed, with high efficiency, and no energy waste.

SAMPLE PROBLEM, CAUSE, EFFECT, AND CORRECTIVE/PREVENTIVE ACTIONS

In relation to previous sections, this section provides sample cases for the main energy consuming equipment in buildings.

Heating, Ventilation and Air Conditioning System

Table 9.1.1 Air Handling Unit

		Equipment Type: Air Handling Units		
		Preventive Maintenance		
Maintenance Inspection	Acceptable Performance	Problem/Effect in case of low performance	Some possible causes in case of low performance - *SAMPLE*	Corrective/Preventive Action
Measure airflow rate (m³/hr)	Current Value = ±5% of Testing and Commissioning Value	- Problem: Low airflow - Effect: Low Cooling in served area	- Increase in Static Pressure - Worn Bearing - Worn Belt - Low Motor Efficiency	- Check any dampers is closed, filters clogged, etc. - Replace Bearing - Replace Belt - Check winding and restore
Check motor's full load current (Amps)	Match value on data plate			
Measure Cooling Coil Performance on Full Load (Air Off Coil Temperature °C)	Current Value = ±5% of Testing and Commissioning Value	- Problem: High off coil temperature - Effect: Low cooling in served area	- Dirty Cooling Coil - Blocked Strainer High supply chilled water temperature	- Clean Coil - Clean Strainer - Check and reduce chilled temperature
Measure Cooling Coil Performance (Pressure Drop, PSI)	Current Value = ±5% of Cooling Coil Selection Pressure Drop	- Problem: High-pressure drop - Effect: Higher electrical consumption in pump's motor	- Scales accumulated inside the coil Increased chilled water flow rate inside the coil than T&C	- Clean or replace Coil - Check DRV & PICV setting and check DPT index value.
Measure Variable Frequency Drive effectiveness	In-voltage is same value of main source voltage, out -voltage = % Frequency.	- Problem: Higher/Lower required power - Effect: higher energy consumption. Less efficiency.	- Drive defected - Meters not calibrated - Ambient temperature - Load defected	- Check drive cooling fans and fix it. - Maintain the ambient temperature to accepted value. - Check meters calibration and recalibrate - Check load parameter and adjust
Measure 2-way/3-way control valves response to space temperature	100% functionality with design intent	- Problem: Stuck open valve - Effect: Overcooling, higher consumption,	- Stuck valve - Open or Close Loosed signal from BMS	- Restore valve to automatic - Restore signal and logic
Measure chilled water temperature difference (Delta-T) °C	100% ± 5% achieving desired Delta-T	- Problem: Low Delta T - Effect: Higher consumption	- Blocked Coils - Stuck Opened Valve Low Setpoints (Temperature or System DPT)	- Clean Coils - Restore valve to auto mode - Adjust set points

		Predictive Maintenance		
Maintenance Task	Acceptable Performance	Problem/Effect in case of low performance	Some possible causes in case of low performance - *SAMPLE*	Corrective/Preventive Action
Measure airflow rate (m³/hr)	Operating value within predefined alarm limits	- Problem: Low airflow - Effect: Low Cooling in served area	- Increase in Static Pressure - Worn Bearing - Worn Belt - Low Motor Efficiency	- Check any dampers is closed, filters clogged, etc. - Replace Bearing - Replace Belt - Check winding and restore
Measure air off coil temperature (°C)	Operating value within predefined alarm limits	- Problem: High off coil temperature - Effect: Low cooling in served area	- Dirty Cooling Coil - Blocked Strainer - High supply chilled water temperature	- Clean Coil - Clean Strainer - Check and reduce chilled temperature
Measure Fan's Motor Power Consumption (kWh)	Operating value within predefined alarm limits	- Problem: High power consumption - Effect: Low cooling in served area	- Increase in Static Pressure - Low Motor Efficiency - Damaged winding	- Check any dampers is closed, filters clogged, etc. - Replace Bearing - Replace Belt - Check winding and restore
Measure chilled water temperature	Operating value within predefined alarm limits	- Problem: Low Delta T	- Blocked Coils - Stuck Opened Valve	- Clean Coils - Restore valve to auto mode

Table 9.1.2 Fan Coil Units

Equipment Type: Fan Coil Units

| | Preventive Maintenance | | | |
Maintenance Inspection	Acceptable Performance	Problem/ Effect in case of low performance	Some possible causes in case of low performance - *SAMPLE*	Corrective/ Preventive Action
Measure airflow rate (m³/hr)	Current Value = ± 5% of Testing and Commissioning Value	- Problem: Low airflow - Effect: Low Cooling in served area	- Increase in Static Pressure - Worn Bearing - Worn Belt - Low Motor Efficiency	- Check any dampers is closed, filters clogged, etc. - Replace Bearing - Replace Belt - Check winding and restore
Check motor's full load current (Amps)	Match value on data plate			
Measure Cooling Coil Performance on Full Load (Air Off Coil Temperature °C)	Current Value = ± 5% of Testing and Commissioning Value	- Problem: High off coil temperature - Effect: Low cooling in served area	- Dirty Cooling Coil - Blocked Strainer - High supply chilled water temperature	- Clean Coil - Clean Strainer - Check and reduce chilled temperature
Measure Cooling Coil Performance (Pressure Drop, PSI)	Current Value = ± 5% of Cooling Coil Selection Pressure Drop	- Problem: High-pressure drop - Effect: Higher electrical consumption in pump's motor	- Scales accumulated inside the coil - Increased chilled water flow rate inside the coil than T&C	- Clean or replace Coil - Check DRV & PICV setting and check DPT index value.
Measure Variable Frequency Drive effectiveness	In-voltage is same value of main source voltage, out –voltage = % Frequency.	- Problem: Higher/Lower required power - Effect: higher energy consumption. Less efficiency	- Drive defected - Meters not calibrated - Ambient temperature - Load defected	- Check drive cooling fans and fix it. - Maintain the ambient temperature to accepted value. - Check meters calibration and recalibrate - Check load parameter and adjust
Measure 2-way/3-way control valves response to space temperature	100% functionality with design intent	- Problem: Stuck open valve - Effect: Overcooling, higher consumption,	- Stuck valve - Open or Close - Loosed signal from BMS	- Restore valve to automatic - Restore signal and logic
Measure chilled water temperature difference (Delta-T) °C	100% ± 5% achieving desired Delta-T	- Problem: Low Delta T - Effect: Higher consumption	- Blocked Coils - Stuck Opened Valve - Low Setpoints (Temperature or System DPT)	- Clean Coils - Restore valve to auto mode - Adjust set points

		Predictive Maintenance		
Maintenance Task	Acceptable Performance	Problem/ Effect in case of low performance	Some possible causes in case of low performance - *SAMPLE*	Corrective/ Preventive Action
Measure airflow rate (m³/hr)	Operating value within predefined alarm limits	- Problem: Low airflow - Effect: Low Cooling in served area	- Increase in Static Pressure - Worn Bearing - Worn Belt - Low Motor Efficiency	- Check any dampers is closed, filters logged, etc. - Replace Bearing - Replace Belt
Measure airflow rate (m³/hr)	Operating value within predefined alarm limits	- Problem: Low airflow - Effect: Low Cooling in served area	- Increase in Static Pressure - Worn Bearing - Worn Belt - Low Motor Efficiency	- Check any dampers is closed, filters logged, etc. - Replace Bearing - Replace Belt - Check winding and restore
Measure air off coil temperature (°C)	Operating value within predefined alarm limits	- Problem: High off coil temperature - Effect: Low cooling in served area	- Dirty Cooling Coil - Blocked Strainer - High supply chilled water temperature	- Clean Coil - Clean Strainer - Check and reduce chilled temperature
Measure Fan's Motor Power Consumption (kWh)	Operating value within predefined alarm limits	- Problem: High power consumption - Effect: Low cooling in served area	- Increase in Static Pressure - Low Motor Efficiency - Damaged winding	- Check any dampers is closed, filters clogged, etc. - Replace Bearing - Replace Belt - Check winding and restore
Measure chilled water temperature difference (Delta-T) °C	Operating value within predefined alarm limits	- Problem: Low Delta T - Effect: Higher consumption	- Blocked Coils - Stuck Opened Valve - Low Setpoints (Temperature or System DhT)	- Clean Coils - Restore valve to auto mode - Adjust set points

Table 9.1.3 Energy Recovery Units (i.e. Heat Wheels)

Equipment Type: Energy Recovery Unit

Preventive Maintenance

Maintenance Inspection	Acceptable Performance	Problem/Effect in case of low performance	Some possible causes in case of low performance - *SAMPLE*	Corrective/Preventive Action
Measure airflow rate (m³/hr)	Current Value = ± 5% of Testing and Commissioning Value	- Problem: Low airflow - Effect: Low Cooling in served area	- Increase in Static Pressure - Worn Bearing - Worn Belt - Low Motor Efficiency	- Check any dampers is closed, filters clogged, etc. - Replace Bearing - Replace Belt - Check winding and restore
Check motor's full load current (Amps)	Match value on data plate			
Measure Energy Recovery Performance on Full Load (Air Off energy recovery unit temperature °C)	Current Value = ± 5% of Testing and Commissioning Value	- Problem: High supply temperature - Effect: insufficient thermal comfort.	- Clogged Coil - Damaged Coil - Dirty Coil	- Clean coil - Repair coil
Measure Cooling Coil Performance (Pressure Drop, PSI)	Current Value = ± 5% of Cooling Coil Selection Pressure Drop	- Problem: High-pressure drop - Effect: Higher electrical consumption in pump's motor	- Scales accumulated inside the coil - Increased chilled water flow rate inside the coil than T&C	- Clean coil - Maintain flow rate as per design

Predictive Maintenance

Maintenance Task	Acceptable Performance	Problem/Effect in case of low performance	Some possible causes in case of low performance - *SAMPLE*	Corrective/Preventive Action
Measure airflow rate (m³/hr)	Operating value within predefined alarm limits	- Problem: Low airflow - Effect: Low Cooling in served area	- Increase in Static Pressure - Worn Bearing - Worn Belt - Low Motor Efficiency	- Check any dampers is closed, filters clogged, etc. - Replace Bearing - Replace Belt - Check winding and restore
Measure air off energy recovery unit temperature (°C)	Operating value within predefined alarm limits	- Problem: High supply temperature - Effect: insufficient thermal comfort.	- Clogged Coil - Damaged Coil - Dirty Coil	- Clean coil - Repair coil
Measure Fan's Motor Power Consumption (kWh)	Operating value within predefined alarm limits	- Problem: High power consumption - Effect: Low cooling in served area	- Increase in Static Pressure - Low Motor Efficiency - Damaged winding	- Check any dampers is closed, filters clogged, etc. - Replace Bearing - Replace Belt - Check winding and restore

Table 9.1.4 Boilers

Equipment Type: Boilers

Preventive Maintenance

Maintenance Inspection	Acceptable Performance	Problem/Effect in case of low performance	Some possible causes in case of low performance - *SAMPLE*	Corrective/Preventive Action
Measure fuel combustion efficiency	± 5% of original efficiency	- Problem: Low Efficiency - Effect: insufficient supply water temperature/high energy consumption	- Air fuel ratio is not balanced - Combustion process is not efficient - Supply of high amount of water, more than design flow rate.	- Test and calibrate air/fuel ratio - Do all required cleaning at burners, heat exchangers, etc. - Supply water flow-rate as per design and check on regular basis.
Inspect steam leakage	No Steam Leakage	- Problem: Steam Leakage - Effect: high energy consumption	- Leakage in tubes, or plates, etc.	- Identify leakage location and restore - Check pressure on regular basis

Predictive Maintenance

Maintenance Task	Acceptable Performance	Problem/Effect in case of low performance	Some possible causes in case of low performance - *SAMPLE*	Corrective/Preventive Action
Outlet Water Temperature	Operating value within predefined alarm limits	- Problem: Low water temperature. - Effect: insufficient supply water temperature/high energy consumption	- Air fuel ratio is not balanced - Combustion process is not efficient - Supply of high amount of water, more than design flow rate.	- Test and calibrate air/fuel ratio - Do all required cleaning at burners, heat exchangers, etc. - Supply water flow-rate as per design and check on regular basis.
Primary System Water Pressure	Operating value within predefined alarm limits	- Problem: Steam Leakage - Effect: high energy consumption	- Leakage in tubes, or plates, etc.	- Identify leakage location and restore - Check pressure on regular basis

Table 9.1.5 Pumps

Equipment Type: Pumps

Preventive Maintenance

Maintenance Inspection	Acceptable Performance	Problem/ Effect in case of low performance	Some possible causes in case of low performance - *SAMPLE*	Corrective/ Preventive Action
Measure full speed water flow rate (gpm, lps)	Match value on data plate	- Problem: Low water flow rate. - Effect: insufficient water delivery – based on function.	- Clogged strainer - Air clogged in the system - Cavitation - Low motor speed - Torn impeller	- Clean strainer - Evacuate air from system - Fix motor winding
Check motor's full load current (Amps)	Match value on data plate	- Problem: Low water flow - Effect: Pump not capable of functioning as designed, high energy consumption	- Increase in the pump head. - Worn Bearing - Blocked strainers - Low Motor Efficiency	- Verify pump head and check impeller - Replace Bearing - Clean strainer - Check winding and restore
Measure Variable Frequency Drive effectiveness	In-voltage is the same value of main source voltage, out –voltage = % Frequency.	- Problem: Higher/Lower required power - Effect: higher energy consumption. Less efficiency	- Drive defected - Meters not calibrated - Ambient temperature - Load defected	- Check drive cooling fans and fix it. - Maintain the ambient temperature to accepted value. - Check meters calibration and recalibrate - Check load parameter and adjust

Predictive Maintenance

Maintenance Task	Acceptable Performance	Problem/ Effect in case of low performance	Some possible causes in case of low performance - *SAMPLE*	Corrective/ Preventive Action
Measure water flow rate (gpm, lps)	Operating value within predefined alarm limits	- Problem: Low water flow rate. - Effect: insufficient water delivery – based on function.	- Clogged strainer - Air clogged in the system - Cavitation - Low motor speed - Torn impeller	- Clean strainer - Evacuate air from system - Fix motor winding
Measure Pump's Motor Power Consumption (kWh)	Operating value within predefined alarm limits	- Problem: kWh more than normal range - Effect: High energy consumption.	- Damaged bearing - Blocked strainer - Damaged PF	- Replace bearing - Clean strainer - Replace PF
Motor running current	Operating value within predefined alarm limits	- Problem: kWh more than normal range - Effect: High energy consumption.	- Damaged bearing - Blocked strainer - Damaged PF	- Replace bearing - Clean strainer - Replace PF

Table 9.1.6 Close Control Units

Equipment Type: Close Control Units

Maintenance Inspection	Acceptable Performance	Preventive Maintenance		Corrective/ Preventive Action
		Problem/ Effect in case of low performance	Some possible causes in case of low performance - *SAMPLE*	
Measure airflow rate (m³/hr)	Current Value = ± 5% of Testing and Commissioning Value	- Problem : Low airflow - Effect: Low Cooling in served area	- Increase in Static Pressure - Worn Bearing - Worn Belt - Low Motor Efficiency	- Check any dampers is closed, filters clogged, etc. - Replace Bearing - Replace Belt - Check winding and restore
Check fan motor's full load current (Amps)	Match value on data plate			
Check DX unit compressor full load current (Amps)	Match value on data plate	- Problem : High Amps - Effect: High energy consumption	- Low refrigerant level - Low Motor Efficiency	- Check and maintain refrigerant level - Check winding and restore
Measure Cooling Coil Performance on Full Load (Air Off Coil Temperature °C)	Current Value = ± 5% of Testing and Commissioning Value	- Problem : High off coil temperature - Effect: Low cooling in served area	- Dirty Cooling Coil - Blocked Strainer - High supply chilled water temperature	- Clean Coil - Clean Strainer - Check and reduce chilled temperature
Measure Cooling Coil Performance (Pressure Drop, PSI)	Current Value = ± 5% of Cooling Coil Selection Pressure Drop	- Problem : High-pressure drop - Effect: Higher electrical consumption in pump's motor	- Scales accumulated inside the coil increased chilled water flow rate inside the coil than T&C	- Clean or replace Coil - Check DRV & PICV setting and check DPT index value.
Measure Variable Frequency Drive effectiveness	In-voltage is same value of main source voltage, out –voltage = % Frequency.	- Problem : Higher/Lower required power - Effect: higher energy consumption. Less efficiency	- Drive defected - Meters not calibrated - Ambient temperature - Load defected	- Check drive cooling fans and fix it. - Maintain the ambient temperature to accepted value. - Check meters' calibration and recalibrate - Check load parameter and adjust
Measure 2-way/3-way control valves response to space temperature	100% functionality with design intent	- Problem : Stuck open valve - Effect: Overcooling, higher consumption,	- Stuck valve - Open or Close Loosed signal from BMS	- Restore valve to automatic - Restore signal and logic
Measure chilled water temperature difference (Delta-T) °C	100%± 5% achieving desired Delta-T	- Problem : Low Delta T - Effect: Higher consumption	- Blocked Coils - Stuck Opened Valve - Low Setpoints (Temperature or System DPT)	- Clean Coils - Restore valve to auto mode - Adjust set points
Measure DX unit Energy Efficiency Ratio EER	± 5% value on data plate	- Problem : lower EER - Effect: Higher energy consumption	- Low refrigerant level - Low Motor Efficiency	- Check and maintain refrigerant level - Check winding and restore

(Continued)

Table 9.1.6 Close Control Units (*Concluded*)

Maintenance Task	Acceptable Performance	Predictive Maintenance		Corrective/ Preventive Action
		Problem/ Effect in case of low performance	Some possible causes in case of low performance - *SAMPLE*	
Measure airflow rate (m³/hr)	Operating value within predefined alarm limits	- Problem: Low airflow - Effect: Low Cooling in served area	- Increase in Static Pressure - Worn Bearing - Worn Belt - Low Motor Efficiency	- Check any dampers is closed, filters clogged, etc. - Replace Bearing - Replace Belt - Check winding and restore
Measure air off coil temperature (°C)	Operating value within predefined alarm limits	- Problem: High off coil temperature - Effect: Low cooling in served area	- Dirty Cooling Coil - Blocked Strainer - High supply chilled water temperature	- Clean Coil - Clean Strainer - Check and reduce chilled temperature
Check DX unit compressor full load current (Amps)	Operating value within predefined alarm limits	- Problem: High Amps - Effect: High energy consumption	- Low refrigerant level - Low Motor Efficiency	- Check and maintain refrigerant level - Check winding and restore
Measure chilled water temperature difference (Delta-T) °C	Operating value within predefined alarm limits	- Problem: Low Delta T - Effect: Higher consumption	- Blocked Coils - Stuck Opened Valve - Low Setpoints (Temperature or System DPT)	- Clean Coils - Restore valve to auto mode - Adjust set points

Table 9.1.7 Fans

Equipment Type: Fans

Preventive Maintenance

Maintenance Inspection	Acceptable Performance	Problem/ Effect in case of low performance	Some possible causes in case of low performance - *SAMPLE*	Corrective/ Preventive Action
Measure airflow rate (m³/hr)	Match value on data plate	- Problem: Low airflow - Effect: Low Cooling in served area	- Increase in Static Pressure - Worn Bearing - Worn Belt - Low Motor Efficiency	- Check any dampers is closed, filters clogged, etc. - Replace Bearing - Replace Belt - Check winding and restore
Check motor's full load current (Amps)	Match value on data plate			
Measure Variable Frequency Drive effectiveness	In-voltage is same value of main source voltage, out -voltage = % Frequency.	- Problem: Higher/Lower required power - Effect: higher energy consumption. Less efficiency	- Drive defected - Meters not calibrated - Ambient temperature - Load defected	- Check drive cooling fans and fix it. - Maintain the ambient temperature to accepted value. - Check meters calibration and recalibrate - Check load parameter and adjust

Predictive Maintenance

Maintenance Task	Acceptable Performance	Problem/ Effect in case of low performance	Some possible causes in case of low performance - *SAMPLE*	Corrective/ Preventive Action
Measure airflow rate (m³/hr)	Operating value within predefined alarm limits	- Problem: Low airflow - Effect: Low Cooling in served area	- Increase in Static Pressure - Worn Bearing - Worn Belt - Low Motor Efficiency	- Check any dampers is closed, filters clogged, etc. - Replace Bearing - Replace Belt - Check winding and restore
Measure Pump's Motor Power Consumption (kWh)	Operating value within predefined alarm limits	- Problem: High power consumption - Effect: Low cooling in served area	- Increase in Static Pressure - Low Motor Efficiency - Damaged winding	- Check any dampers is closed, filters clogged, etc. - Replace Bearing - Replace Belt - Check winding and restore

Table 9.1.8 Cooling Towers

Equipment Type: Cooling Towers

Preventive Maintenance

Maintenance Inspection	Acceptable Performance	Problem/Effect in case of low performance	Some possible causes in case of low performance - *SAMPLE*	Corrective/Preventive Action
Check fan motor's full load current (Amps)	Match value on data plate	- Problem: High Amps - Effect: High consumption, low airflow rate extraction	- Blocked louvers - Worn bearing - Damaged winding	- Clean louvers - Replace bearing - Fix winding
Cooling Tower Range (In-Out Water Temperature) – Full Load (°C)	Match value on data plate	- Problem: Low water range - Effect: High outlet temperature.	- Blocked louvers - Higher amount of water discharged from sprinklers - Damaged fan blades	- Clean and fix louvers - Check and calibrate water from sprinklers - Check and replace fan blades
Measure Variable Frequency Drive effectiveness	In-voltage is the same value of main source voltage. out –voltage = % Frequency.	- Problem: Higher/Lower required power - Effect: higher energy consumption. Less efficiency	- Drive defected - Meters not calibrated - Ambient temperature - Load defected	- Check drive cooling fans and fix it. - Maintain the ambient temperature to accepted value. - Check meters calibration and recalibrate - Check load parameter and adjust

Predictive Maintenance

Maintenance Task	Acceptable Performance	Problem/Effect in case of low performance	Some possible causes in case of low performance - *SAMPLE*	Corrective/Preventive Action
Check fan motor's current (Amps)	Operating value within predefined alarm limits	- Problem: High Amps - Effect: High consumption, low airflow rate extraction	- Blocked louvers - Worn bearing - Damaged winding	- Clean louvers - Replace bearing - Fix winding
Cooling Tower Range (In-Out Water Temperature) – Full Load (°C)	Operating value within predefined alarm limits	- Problem: Low water range - Effect: High outlet temperature.	- Blocked louvers - Higher amount of water discharged from sprinklers - Damaged fan blades	- Clean and fix louvers - Check and calibrate water from sprinklers - Check and replace fan blades

Table 9.1.9 Air Cooled Chillers

Equipment Type: Air Cooled Chillers

Maintenance Inspection	Acceptable Performance	Preventive Maintenance		Corrective/Preventive Action
		Problem/Effect in case of low performance	Some possible causes in case of low performance - *SAMPLE*	
Check compressor full-load current (Amps)	Match value on data plate	- Problem: High Amps - Effect: High energy consumption	- Low refrigerant level - Low Motor Efficiency	- Check and maintain refrigerant level - Check winding and restore
Measure Variable Frequency Drive effectiveness	In-voltage is the same value of main source voltage, out -voltage = % Frequency.	- Problem: Higher/Lower required power - Effect: higher energy consumption. Less efficiency	- Drive defected - Meters not calibrated - Ambient temperature - Load defected	- Check drive cooling fans and fix it. - Maintain the ambient temperature to accepted value. - Check meters calibration and recalibrate - Check load parameter and adjust
Check Condenser Fan motor's full load current (Amps)	Match value on data plate	- Problem: High Amps - Effect: High energy consumption, low airflow rate extraction	- Blocked condenser coil - Worn bearing - Damaged winding	- Clean condenser coil - Replace bearing - Fix winding
Evaporator pressure drop (PSI)	Match value on data plate	- Problem: High-pressure drop - Effect: Higher electrical consumption in pump's motor	- Scales accumulated inside the evaporator. - Increased chilled water flow rate inside the coil than T&C	- Clean evaporator - Check DRV setting and adjust to limit flow as per T&C data.
Refrigerant leaks test	No Leak	- Problem: High Amps - Effect: High energy consumption, low airflow rate extraction	- Leakage in direction expansion cycle	- Detect leak and ensure direct expansion cycle has no leakage

(Continued)

Table 9.1.9 Air Cooled Chillers (*Concluded*)

Maintenance Task	Predictive Maintenance			
	Acceptable Performance	Problem/Effect in case of low performance	Some possible causes in case of low performance - *SAMPLE*	Corrective/Preventive Action
Check Compressor motor's current (Amps)	Operating value within predefined alarm limits	- Problem: High Amps - Effect: High energy consumption	- Low refrigerant level - Low Motor Efficiency	- Check and maintain refrigerant level - Check winding and restore
Chilled water supply temperature (°C)	Operating value within predefined alarm limits	- Problem: High Supply Temperature - Effect: Low cooling in facility	- Ineffective heat exchange on evaporator - Issues with expansion valve - Refrigerant leak - No sufficient heat exchange on	- Clean evaporator - Check and replace expansion valve based on enthalpy value - Detect leak and ensure direct expansion cycle has no leakage
Chilled water supply temperature (°C)	Operating value within predefined alarm limits	- Problem: High Supply Temperature - Effect: Low cooling in facility	- Ineffective heat exchange on evaporator - Issues with expansion valve - Refrigerant leak - No sufficient heat exchange on consider	- Clean evaporator - Check and replace expansion valve based on enthalpy value - Detect leak and ensure direct expansion cycle has no leakage - Clean condenser coil, check condensing temperature and condenser effectiveness

Table 9.1.10 Heat Exchangers

Equipment Type: Heat Exchangers

Preventive Maintenance

Maintenance Inspection	Acceptable Performance	Problem/Effect in case of low performance	Some possible causes in case of low performance - *SAMPLE*	Corrective/Preventive Action
Measure pressure drop (PSI)	Match value on data plate	- Problem: High pressure drop - Effect: Higher electrical consumption in pump's motor	- Scales accumulated inside the plates. - Increased chilled water flow rate inside the plates than T&C	- Clean plates - Check DRV setting and adjust to limit flow as per T&C data.
Heat exchanger effectiveness (%) - (In/Out Temperature)	Match value on data plate	- Problem: High water supply temperature - Effect: less cooling, depends on the function.	- Scales accumulated inside the plates. - Increased chilled water flow rate inside the plates than T&C - Higher In water temperature than T&C values.	- Clean plates - Check DRV setting and adjust to limit flow as per T&C data. - Investigate reasons of high In water temperature and rectify.

Predictive Maintenance

Maintenance Task	Acceptable Performance	Problem/Effect in case of low performance	Some possible causes in case of low performance - *SAMPLE*	Corrective/Preventive Action
Chilled water supply temperature (°C)	Operating value within predefined alarm limits	- Problem: High water supply temperature - Effect: less cooling, depends on the function.	- Scales accumulated inside the plates. - Increased chilled water flow rate inside the plates than T&C - Higher In water temperature than T&C values.	- Clean plates - Check DRV setting and adjust to limit flow as per T&C data. - Investigate reasons of high In water temperature and rectify.

Table 9.1.11 Water Cooled Chillers

Equipment Type: Water Cooled Chillers

Maintenance Inspection	Acceptable Performance	Preventive Maintenance		Corrective/ Preventive Action
		Problem/ Effect in case of low performance	Some possible causes in case of low performance - *SAMPLE*	
Check compressor full-load current (Amps)	Match value on data plate	- Problem: High Amps - Effect: High energy consumption	- Low refrigerant level - Low Motor Efficiency	- Check and maintain refrigerant level - Check winding and restore
Evaporator pressure drop (PSI)	Match value on data plate	- Problem: High pressure drop - Effect: Higher electrical consumption in pump's motor	- Scales accumulated inside the evaporator. - Increased chilled water flow rate inside the coil than T&C	- Clean evaporator - Check DRV setting and adjust to limit flow as per T&C data.
Measure Variable Frequency Drive effectiveness	In-voltage is same value of main source voltage. out –voltage = % Frequency.	- Problem: Higher/Lower required power - Effect: higher energy consumption. Less efficiency	- Drive defected - Meters not calibrated - Ambient temperature - Load defected	- Check drive cooling fans and fix it. - Maintain the ambient temperature to accepted value. - Check meters calibration and recalibrate - Check load parameter and adjust
Condenser pressure drop (PSI)	Match value on data plate	- Problem: High pressure drop - Effect: Higher electrical consumption in pump's motor	- Scales accumulated inside the condenser. - Increased chilled water flow rate inside the coil than T&C	- Clean condenser - Check DRV setting and adjust to limit flow as per T&C data.
Refrigerant leaks test	No Leak	- Problem: High Amps - Effect: High energy consumption, low airflow rate extraction	- Leakage in direction expansion cycle	- Detect leak and ensure direct expansion cycle has no leakage

Maintenance Task	Predictive Maintenance			Corrective/ Preventive Action
	Acceptable Performance	Problem/ Effect in case of low performance	Some possible causes in case of low performance - *SAMPLE*	
Check Compressor motor's current (Amps)	Operating value within predefined alarm limits	- Problem: High Amps - Effect: High energy consumption	- Low refrigerant level - Low Motor Efficiency	- Check and maintain refrigerant level - Check winding and restore
Chilled water supply temperature (°C)	Operating value within predefined alarm limits	- Operating value within predefined alarm limits	- Problem: High Supply Temperature - Effect: Low cooling in facility	- Ineffective heat exchange on evaporator - Issues with expansion valve - Refrigerant leak - No sufficient heat exchange on consider
Operating pressure (PSI)	Operating value within predefined alarm limits	- Operating value within predefined alarm limits	- Problem: High pressure drop Effect: Higher electrical consumption in pump's motor	- Scales accumulated inside the evaporator. Increased chilled water flow rate inside the coil than T&C

Table 9.1.12 Direct Expansion Air Conditioners

Equipment Type: Direct Expansion Air Conditioners

Preventive Maintenance

Maintenance Inspection	Acceptable Performance	Problem/ Effect in case of low performance	Some possible causes in case of low performance - *SAMPLE*	Corrective/ Preventive Action
Check compressor full-load current (Amps)	Match value on data plate	- Problem: High Amps - Effect: High energy consumption	- Low refrigerant level - Low Motor Efficiency	- Check and maintain refrigerant level - Check winding and restore
Refrigerant leaks test	No Leak	- Problem: High Amps - Effect: High energy consumption, low airflow rate extraction	- Leakage in direction expansion cycle	- Detect leak and ensure direct expansion cycle has no leakage

Predictive Maintenance

Maintenance Task	Acceptable Performance	Problem/ Effect in case of low performance	Some possible causes in case of low performance - *SAMPLE*	Corrective/ Preventive Action
Check Compressor motor's current (Amps)	Operating value within predefined alarm limits	- Problem: High Amps - Effect: High energy consumption	- Low refrigerant level - Low Motor Efficiency	- Check and maintain refrigerant level - Check winding and restore

Table 9.1.13 Economizers

Equipment Type: Economizers

Preventive Maintenance

Maintenance Inspection	Acceptable Performance	Problem/Effect in case of low performance	Some possible causes in case of low performance - *SAMPLE*	Corrective/Preventive Action
Supply air flow rate after mixing (m³/hr)	Match value on data plate	- Problem: Low airflow rate - Effect: Depends on the application. If used for free cooling may result in low cooling inside the building.	- Stuck dampers - Blocked air intakes	- Check auto function of dampers - Check air intakes for any blockage and clean
Supply air flow rate temperature after mixing (°C)	Match value on data plate	- Problem: High supply air temperature - Effect: Depends on the application. If used for free cooling may result in low cooling inside the building.	- Air mixing ratio not accurate, more return air is mixing with outside air	- Air flow rate requires re-balancing by adjusting dampers settings.

Predictive Maintenance

Maintenance Task	Acceptable Performance	Problem/Effect in case of low performance	Some possible causes in case of low performance - *SAMPLE*	Corrective/Preventive Action
Supply air flow rate after mixing (m³/hr)	Match value on data plate	- Problem: Low airflow rate - Effect: Depends on the application. If used for free cooling may result in low cooling inside the building.	- Stuck dampers - Blocked air intakes	- Check auto function of dampers - Check air intakes for any blockage and clean
Supply air flow rate temperature after mixing (°C)	Match value on data plate	- Problem: High supply air temperature - Effect: Depends on the application. If used for free cooling may result in low cooling inside the building.	- Air mixing ratio not accurate, more return air is mixing with outside air	- Air flow rate requires re-balancing by adjusting dampers settings.

Table 9.1.14 Air Compressors

Equipment Type: Air Compressors

Maintenance Inspection	Preventive Maintenance			Corrective/Preventive Action
	Acceptable Performance	Problem/Effect in case of low performance	Some possible causes in case of low performance - *SAMPLE*	
Produced air pressure (PSI)	Match value on data plate	- Problem: Low Produced Air Pressure - Effect: Impact on operational function	- Air Leakage - Compressors degrading power	- Check and rectify any air leakage found - Check compressor effectiveness and rectify

Water Supply System

Table 9.1.15 Domestic Water Pump Set, Irrigation Pump, and Water Features Pumps

Equipment Type: Pumps

Preventive Maintenance

Maintenance Inspection	Acceptable Performance	Problem/Effect in case of low performance	Some possible causes in case of low performance - *SAMPLE*	Corrective/Preventive Action
Measure full speed water flow rate (gpm, lps)	Match value on data plate	Problem: Low water flow rate. Effect: insufficient water delivery – based on function.	- Clogged strainer - Air clogged in the system - Cavitation - Low motor speed - Torn impeller	- Clean strainer - Evacuate air from system - Fix motor winding
Check motor's full load current (Amps)	Match value on data plate	Problem: Low water flow Effect: Pump not capable of functioning as designed, high energy consumption	- Increase in the pump head. - Worn Bearing - Blocked strainers - Low Motor Efficiency	- Verify pump head and check impeller - Replace Bearing - Clean strainer - Check winding and restore
Measure Variable Frequency Drive effectiveness	In-voltage is the same value of main source voltage, out –voltage = % Frequency.	Problem: Higher/Lower required power Effect: higher energy consumption. Less efficiency	- Drive defected - Meters not calibrated - Ambient temperature - Load defected	- Check drive cooling fans and fix it. - Maintain the ambient temperature to accepted value. - Check meters calibration and recalibrate - Check load parameter and adjust

Predictive Maintenance

Maintenance Task	Acceptable Performance	Problem/Effect in case of low performance	Some possible causes in case of low performance - *SAMPLE*	Corrective/Preventive Action
Measure water flow rate (gpm, lps)	Operating value within predefined alarm limits	Problem: Low water flow rate. Effect: insufficient water delivery – based on function.	- Clogged strainer - Air clogged in the system - Cavitation - Low motor speed	- Clean strainer - Evacuate air from system - Fix motor winding
Measure Pump's Motor Power Consumption (kWh)	Operating value within predefined alarm limits	Problem: kWh more than normal range Effect: High energy consumption.	- Damaged bearing - Blocked strainer - Damaged PF	- Replace bearing - Clean strainer - Replace PF
Motor running current	Operating value within predefined alarm limits	Problem: kWh more than normal range Effect: High energy consumption.	- Damaged bearing - Blocked strainer - Damaged PF	- Replace bearing - Clean strainer - Replace PF

Table 9.1.16 Heat Exchangers

Equipment Type: Heat Exchangers

Preventive Maintenance

Maintenance Inspection	Acceptable Performance	Problem/ Effect in case of low performance	Some possible causes in case of low performance - *SAMPLE*	Corrective/ Preventive Action
Measure pressure drop (PSI)	Match value on data plate	- Problem: High pressure drop - Effect: Higher electrical consumption in pump's motor	- Scales accumulated inside the plates. - Increased chilled water flow rate inside the plates than T&C	- Clean plates - Check DRV setting and adjust to limit flow as per T&C data.
Heat exchanger effectiveness (%) - (In/ Out Temperature)	Match value on data plate	- Problem: High water supply temperature - Effect: less cooling, depends on the function.	- Scales accumulated inside the plates. - Increased chilled water flow rate inside the plates than T&C - Higher In water temperature than T&C values.	- Clean plates - Check DRV setting and adjust to limit flow as per T&C data. - Investigate reasons of high In water temperature and rectify.

Predictive Maintenance

Maintenance Task	Acceptable Performance	Problem/ Effect in case of low performance	Some possible causes in case of low performance - *SAMPLE*	Corrective/ Preventive Action
Chilled water supply temperature (°C)	Operating value within predefined alarm limits	- Problem: High water supply temperature - Effect: less cooling, depends on the function.	- Scales accumulated inside the plates. - Increased chilled water flow rate inside the plates than T&C - Higher In water temperature than T&C values.	- Clean plates - Check DRV setting and adjust to limit flow as per T&C data. - Investigate reasons of high In water temperature and rectify.

Table 9.1.17 Pressure Reducing Valve Station

Equipment Type: PRV Station

Maintenance Inspection	Preventive Maintenance			Corrective/ Preventive Action
	Acceptable Performance	Problem/ Effect in case of low performance	Some possible causes in case of low performance - *SAMPLE*	
Measure On/ Off Pressure on critical PRVs	Achieve pressure value as per T&C	- Problem: Higher (or lower) water pressure than T&C - Effect: Increase in water consumption in case of higher water pressure, or dissatisfaction of building occupants in case of low water pressure.	- Damaged PRV or change in settings during its operating life. - Dirty screen inside PRV leads to low outlet pressure	- Check settings and recalibrate when required. - Clean screen and recalibrate

- Note: only main PRVs should be recalibrated. For example, recalibrate PRVs main on domestic water lines entering floors

Table 9.1.18 Boilers

Equipment Type: Boilers

Preventive Maintenance

Maintenance Inspection	Acceptable Performance	Problem/Effect in case of low performance	Some possible causes in case of low performance - *SAMPLE*	Corrective/Preventive Action
Measure fuel combustion efficiency	± 5% of original efficiency	- Problem: Low Efficiency - Effect: insufficient supply water temperature/high energy consumption	- Air fuel ratio is not balanced - Combustion process is not efficient - Supply of high amount of water, more than design flow rate.	- Test and calibrate air/fuel ratio - Do all required cleaning at burners, heat exchangers, etc. - Supply water flow-rate as per design and check on regular basis.
Inspect steam leakage	No Steam Leakage	- Problem: Steam Leakage - Effect: high energy consumption	- Leakage in tubes, or plates, etc.	- Identify leakage location and restore - Check pressure on regular basis

Predictive Maintenance

Maintenance Task	Acceptable Performance	Problem/Effect in case of low performance	Some possible causes in case of low performance - *SAMPLE*	Corrective/Preventive Action
Outlet Water Temperature	Operating value within predefined alarm limits	- Problem: Low water temperature. - Effect: insufficient supply water temperature/high energy consumption	- Air fuel ratio is not balanced - Combustion process is not efficient - Supply of high amount of water, more than design flow rate.	- Test and calibrate air/fuel ratio - Do all required cleaning at burners, heat exchangers, etc. - Supply water flow-rate as per design and check on regular basis.
Primary System Water Pressure	Operating value within predefined alarm limits	- Problem: Steam Leakage - Effect: high energy consumption	- Leakage in tubes, or plates, etc.	- Identify leakage location and restore - Check pressure on regular basis

Drainage System

Table 9.1.19 Sump Pumps

Equipment Type: Pumps

Preventive Maintenance

Maintenance Inspection	Acceptable Performance	Problem/ Effect in case of low performance	Some possible causes in case of low performance - *SAMPLE*	Corrective/ Preventive Action
Measure full speed water flow rate (gpm, lps)	Match value on data plate	- Problem: Low water flow rate. - Effect: insufficient water delivery – based on function.	- Clogged strainer - Air clogged in the system - Cavitation - Low motor speed - Torn impeller	- Clean strainer - Evacuate air from system - Fix motor winding
Check motor's full load current (Amps)	Match value on data plate	- Problem: Low water flow - Effect: Pump not capable of functioning as designed, high energy consumption	- Increase in the pump head. - Worn Bearing - Blocked strainers - Low Motor Efficiency	- Verify pump head and check impeller - Replace Bearing - Clean strainer - Check winding and restore
Measure Variable Frequency Drive effectiveness	In-voltage is the same value of main source voltage, out –voltage = % Frequency.	- Problem: Higher/Lower required power - Effect: higher energy consumption. Less efficiency	- Drive defected - Meters not calibrated - Ambient temperature - Load defected	- Check drive cooling fans and fix it. - Maintain the ambient temperature to accepted value. - Check meters calibration and recalibrate - Check load parameter and adjust

Predictive Maintenance

Maintenance Task	Acceptable Performance	Problem/ Effect in case of low performance	Some possible causes in case of low performance - *SAMPLE*	Corrective/ Preventive Action
Measure water flow rate (gpm, lps)	Operating value within predefined alarm limits	- Problem: Low water flow rate. - Effect: insufficient water delivery – based on function.	- Clogged strainer - Air clogged in the system - Cavitation - Low motor speed	- Clean strainer - Evacuate air from system - Fix motor winding
Measure Pump's Motor Power Consumption (kWh)	Operating value within predefined alarm limits	- Problem: kWh more than normal range - Effect: High energy consumption.	- Damaged bearing - Blocked strainer - Damaged PF	- Replace bearing - Clean strainer - Replace PF
Motor running current	Operating value within predefined alarm limits	- Problem: kWh more than normal range - Effect: High energy consumption.	- Damaged bearing - Blocked strainer - Damaged PF	- Replace bearing - Clean strainer - Replace PF

Storm Water Management System

Table 9.1.20 Rain Water Pumps

Equipment Type: Pumps

Preventive Maintenance

Maintenance Inspection	Acceptable Performance	Problem/Effect in case of low performance	Some possible causes in case of low performance - *SAMPLE*	Corrective/Preventive Action
Measure full speed water flow rate (gpm, lps)	Match value on data plate	Problem: Low water flow rate. Effect: insufficient water delivery – based on function.	- Clogged strainer - Air clogged in the system - Cavitation - Low motor speed - Torn impeller	- Clean strainer - Evacuate air from system - Fix motor winding
Check motor's full load current (Amps)	Match value on data plate	Problem: Low water flow Effect: Pump not capable of functioning as designed, high energy consumption	- Increase in the pump head. - Worn Bearing - Blocked strainers - Low Motor Efficiency	- Verify pump head and check impeller - Replace Bearing - Clean strainer - Check winding and restore - Check drive cooling fans and fix it.
Measure Variable Frequency Drive effectiveness	In-voltage is the same value of main source voltage, out –voltage = % Frequency.	Problem: Higher/Lower required power - Effect: higher energy consumption. Less efficiency	- Drive defected - Meters not calibrated - Ambient temperature - Load defected	- Maintain the ambient temperature to accepted value. - Check meters calibration and recalibrate - Check load parameter and adjust

Predictive Maintenance

Maintenance Task	Acceptable Performance	Problem/Effect in case of low performance	Some possible causes in case of low performance - *SAMPLE*	Corrective/Preventive Action
Measure water flow rate (gpm, lps)	Operating value within predefined alarm limits	Problem: Low water flow rate. Effect: insufficient water delivery – based on function.	- Clogged strainer - Air clogged in the system - Cavitation - Low motor speed	- Clean strainer - Evacuate air from system - Fix motor winding
Measure Pump's Motor Power Consumption (kWh)	Operating value within predefined alarm limits	Problem: kWh more than normal range - Effect: High energy consumption.	- Damaged bearing - Blocked strainer - Damaged PF	- Replace bearing - Clean strainer - Replace PF
Motor running current	Operating value within predefined alarm limits	Problem: kWh more than normal range - Effect: High energy consumption.	- Damaged bearing - Blocked strainer - Damaged PF	- Replace bearing - Clean strainer - Replace PF

Building Transportation System

Table 9.1.21 Travelators and Escalators

Equipment Type: Travelators and Escalators				
Preventive Maintenance				
Maintenance Inspection	Acceptable Performance	Problem/ Effect in case of low performance	Some possible causes in case of low performance - *SAMPLE*	Corrective/ Preventive Action
Check motor's full load current (Amps)	Match value on data plate	- Problem: High Amps - Effect: Higher energy consumption	- Low Motor Efficiency - Damaged winding - High operating temperature	- Check motor efficiency and repair - Repair or replace motor winding - Maintain acceptable temperature in surrounding environment
Auto Start/ Stop Command	Auto Start/Stop to be workable	- Problem: Motor continuously working - Effect: Higher energy consumption	- Damage On/Off Sensors - Control logic stuck on (On mode).	- Check functionality of sensor and replace defected ones - Check control logic for Auto Start/Stop command
Predictive Maintenance				
Maintenance Task	Acceptable Performance	Problem/ Effect in case of low performance	Some possible causes in case of low performance - *SAMPLE*	Corrective/ Preventive Action
Measure Motor Power Consumption (kWh)	Operating value within predefined alarm limits	- Problem: High Amps - Effect: Higher energy consumption	- Low Motor Efficiency - Damaged winding - High operating temperature	- Check motor efficiency and repair - Repair or replace motor winding - Maintain acceptable temperature in surrounding environment

Table 9.1.22 Elevators

Equipment Type: Elevators

Preventive Maintenance

Maintenance Inspection	Acceptable Performance	Problem/Effect in case of low performance	Some possible causes in case of low performance - *SAMPLE*	Corrective/Preventive Action
Check motor's full load current (Amps)	Match value on data plate	- Problem: High Amps - Effect: Higher energy consumption	- Low Motor Efficiency - Damaged winding - High operating temperature	- Check motor efficiency and repair - Repair or replace motor winding - Maintain acceptable temperature in surrounding environment
Auto Start/ Stop Command	Auto Start/Stop to be workable	- Problem: Motor continuously working - Effect: Higher energy consumption	- Damage On/Off Sensors - Control logic stuck on (On mode).	- Check functionality of sensor and replace defected ones - Check control logic for Auto Start/Stop command

Predictive Maintenance

Maintenance Task	Acceptable Performance	Problem/Effect in case of low performance	Some possible causes in case of low performance - *SAMPLE*	Corrective/Preventive Action
Measure Motor Power Consumption (kWh)	Operating value within predefined alarm limits	- Problem: High Amps - Effect: Higher energy consumption	- Low Motor Efficiency - Damaged winding - High operating temperature	- Check motor efficiency and repair - Repair or replace motor winding - Maintain acceptable temperature in surrounding environment

Fire Fighting System

Table 9.1.23 Fire Pumps

Equipment Type: Pumps

Preventive Maintenance

Maintenance Inspection	Acceptable Performance	Problem/ Effect in case of low performance	Some possible causes in case of low performance - *SAMPLE*	Corrective/ Preventive Action
Measure full speed water flow rate (gpm, lps)	Match value on data plate	- Problem: Low water flow rate. - Effect: insufficient water delivery - based on function.	- Clogged strainer - Air clogged in the system - Cavitation - Low motor speed - Torn impeller	- Clean strainer - Evacuate air from system - Fix motor winding
Check motor's full load current (Amps)	Match value on data plate	- Problem: Low water flow - Effect: Pump not capable of functioning as designed, high energy consumption	- Increase in the pump head. - Worn Bearing - Blocked strainers - Low Motor Efficiency	- Verify pump head and check impeller - Replace Bearing - Clean strainer - Check winding and restore - Check drive cooling fans and fix it.
Measure Variable Frequency Drive effectiveness	In-voltage is the same value of main source voltage, out –voltage = % Frequency.	- Problem: Higher/Lower required power - Effect: higher energy consumption. Less efficiency	- Drive defected - Meters not calibrated - Ambient temperature - Load defected	- Maintain the ambient temperature to accepted value. - Check meters calibration and recalibrate - Check load parameter and adjust

Predictive Maintenance

Maintenance Task	Acceptable Performance	Problem/ Effect in case of low performance	Some possible causes in case of low performance - *SAMPLE*	Corrective/ Preventive Action
Measure water flow rate (gpm, lps)	Operating value within predefined alarm limits	- Problem: Low water flow rate. - Effect: insufficient water delivery - based on function.	- Clogged strainer - Air clogged in the system - Cavitation - Low motor speed	- Clean strainer - Evacuate air from system - Fix motor winding
Measure Pump's Motor Power Consumption (kWh)	Operating value within predefined alarm limits	- Problem: kWh more than normal range - Effect: High energy consumption.	- Damaged bearing - Blocked strainer - Damaged PF	- Replace bearing - Clean strainer - Replace PF
Motor running current	Operating value within predefined alarm limits	- Problem: kWh more than normal range - Effect: High energy consumption.	- Damaged bearing - Blocked strainer - Damaged PF	- Replace bearing - Clean strainer - Replace PF

Electrical System

Table 9.1.24 Motor Control Centre

Equipment Type: MCC – Starters

Maintenance Inspection	Preventive Maintenance			Corrective/ Preventive Action
	Acceptable Performance	Problem/ Effect in case of low performance	Some possible causes in case of low performance - *SAMPLE*	
In/out voltage at constant frequency	In-voltage is the same value of main source voltage, out −voltage = % Frequency.	- Problem: Higher/Lower required power - Effect: higher energy consumption. Less efficiency	- Drive defected - Meters not calibrated - Ambient temperature - Load defected	- Check drive cooling fans and fix it. - Maintain the ambient temperature to accepted value. - Check meters calibration and recalibrate - Check load parameter and adjust
In/out current at constant frequency	In-current = out current +2%	- Problem: Higher/Lower required power - Effect: higher energy consumption. Less efficiency	- Drive defected - Meters not calibrated - Ambient temperature - Load defected	- Check drive cooling fans and fix it. - Maintain the ambient temperature to accepted value. - Check meters calibration and recalibrate - Check load parameter and adjust
Operating temperature	In accordance with manufacturer recommendations	- Problem: Operating temperature increase lead to drive defect - Effect: more energy consumption	- Drive defected	- Maintain the ambient temperature to accepted value.

Maintenance Task	Acceptable Performance	Predictive Maintenance		Corrective/Preventive Action
		Problem/ Effect in case of low performance	Some possible causes in case of low performance - *SAMPLE*	
Verify the out voltage with frequency	In-voltage is the same value of main source voltage, out –voltage = %	- Problem: Higher/Lower required power - Effect: higher energy consumption. Less efficiency	- Drive defected - Meters not calibrated - Ambient temperature - Load defected	- Check drive cooling fans and fix it. - Maintain the ambient temperature to accepted value.
Verify the out voltage with frequency	In-voltage is the same value of main source voltage, out –voltage = % Frequency.	- Problem: Higher/Lower required power - Effect: higher energy consumption. Less efficiency	- Drive defected - Meters not calibrated - Ambient temperature - Load defected	- Check drive cooling fans and fix it. - Maintain the ambient temperature to accepted value. - Check meters calibration and recalibrate - Check load parameter and

Table 9.1.25 Variable Frequency Drive

Equipment Type: VFD

Preventive Maintenance

Maintenance Inspection	Acceptable Performance	Problem/ Effect in case of low performance	Some possible causes in case of low performance - *SAMPLE*	Corrective/ Preventive Action
In/out voltage at constant frequency	In-voltage is the same value of main source voltage, out –voltage = % Frequency.	- Problem: Higher/Lower required power - Effect: higher energy consumption. Less efficiency	- Drive defected - Meters not calibrated - Ambient temperature - Load defected	- Check drive cooling fans and fix it. - Maintain the ambient temperature to accepted value. - Check meters calibration and recalibrate - Check load parameter and adjust
In/out current at constant frequency	In-current = out current +2%	- Problem: Higher/Lower required power - Effect: higher energy consumption. Less efficiency	- Drive defected - Meters not calibrated - Ambient temperature - Load defected	- Check drive cooling fans and fix it. - Maintain the ambient temperature to accepted value. - Check meters calibration and recalibrate - Check load parameter and adjust
Operating temperature	In accordance with manufacturer recommendations	- Problem: Operating temperature increase lead to drive defect - Effect: more energy consumption	- Drive defected	- Maintain the ambient temperature to accepted value.

Predictive Maintenance

Maintenance Task	Acceptable Performance	Problem/ Effect in case of low performance	Some possible causes in case of low performance - *SAMPLE*	Corrective/ Preventive Action
Verify the out voltage with frequency	In-voltage is the same value of main source voltage, out –voltage = % Frequency.	- Problem: Higher/Lower required power - Effect: higher energy consumption. Less efficiency	- Drive defected - Meters not calibrated - Ambient temperature - Load defected	- Check drive cooling fans and fix it. - Maintain the ambient temperature to accepted value. - Check meters calibration and recalibrate
Verify the out voltage with frequency	In-voltage is the same value of main source voltage, out –voltage = % Frequency.	- Problem: Higher/Lower required power - Effect: higher energy consumption. Less efficiency	- Drive defected - Meters not calibrated - Ambient temperature - Load defected	- Check drive cooling fans and fix it. - Maintain the ambient temperature to accepted value. - Check meters calibration and

Table 9.1.26 Lighting Bulbs

Equipment Type: PRV Station

Maintenance Inspection	Acceptable Performance	Preventive Maintenance		Corrective/ Preventive Action
		Problem/ Effect in case of low performance	Some possible causes in case of low performance - *SAMPLE*	
Measure Lux Level (foot-candle)	Match value on manufacture catalog	- Problem: Low Lux Level - Effect: Low illumination.	- Dirty Light Bulb	- Clean Bulbs on regular basis

Building Management System

Table 9.1.27 Two-way Control Valve

Equipment Type: 2 Way Control Valve

Preventive Maintenance

Maintenance Inspection	Acceptable Performance	Problem/Effect in case of low performance	Some possible causes in case of low performance - *SAMPLE*	Corrective/Preventive Action
Check actuator response in line with control signal	Linear response to control signal on site (0-10 Vdc = 0-100% valve closing/opening)	- Problem: Valve stuck and not regulating according to demand. - Effect: Over cooling/Higher chilled water consumption	- Mechanical fault, valve stuck on open position - Lost signal from BMS - Faulty logic	- Check valve position and return to auto mode - Check valve response to BMS signal on regular basis - Check valve response to control logic
Check feedback correspondence to control signal	Response received on frontend software	- Problem: Command is not received on-site equipment. - Effect: Over cooling/Higher chilled water consumption	- Mechanical fault, valve stuck on open position - Lost signal from BMS - Faulty logic	- Check valve position and return to auto mode - Check valve response to BMS signal on regular basis - Check valve response to control logic
Check response time with respect to command	Closing/Opening time as per original logic (response time may be adjusted for more efficient response)	- Problem: Valve closing/opening response is slow - Effect: Overcooling or/Higher chilled water consumption - Effect: Insufficient cooling/Inconvenient space temperature and humidity	- Faulty logic	- Correct logic and check opening/closing speed

Predictive Maintenance

Maintenance Task	Acceptable Performance	Problem/Effect in case of low performance	Some possible causes in case of low performance - *SAMPLE*	Corrective/Preventive Action
Valves response to AHU/FCU temperature set point	Linear response to set points	- Problem: Valve closing/opening response is slow - Effect: Overcooling or/Higher chilled water consumption - Effect: Insufficient cooling/Inconvenient space temperature and humidity	- Faulty logic	- Correct logic and check opening/closing speed

Table 9.1.28 Differential Pressure Switch

Equipment Type: Differential Pressure Switch				
Preventive Maintenance				
Maintenance Inspection	Acceptable Performance	Problem/ Effect in case of low performance	Some possible causes in case of low performance - *SAMPLE*	Corrective/ Preventive Action
Check DPS functionality with respect to control signal	Linear response to control signal on site/ True signal based on site condition in fans	- Problem: Fans speed is not response to DPS value. - Effect: Higher energy consumption	- Mechanical fault, fan operating on manual speed. - Lost signal from BMS - Faulty logic	- Return fan to Auto mode - Check fan speed response to DPS value and control logic on regular basis
Conduct continuity loop testing	Command/ Response received on frontend software and on-site equipment	- Problem: Command is not received on-site equipment - Effect: Higher energy consumption	- Mechanical fault, fan operating on manual speed. - Lost signal from BMS - Faulty logic	- Return fan to Auto mode - Check fan speed response to DPS value and control logic on regular basis
Compare differential pressure value on frontend and on-site	Command/ Response received on frontend software and on-site equipment	- Problem: Different DPS reading on frontend and on-site - Effect: Higher energy consumption	- Faulty DPS	- Replace DPS, or recalibrate if possible.
Predictive Maintenance				
Maintenance Task	Acceptable Performance	Problem/ Effect in case of low performance	Some possible causes in case of low performance - *SAMPLE*	Corrective/ Preventive Action
Differential pressure value	Value within predefined limits	- Problem: DPS reading outside acceptable limits - Effect: Higher energy consumption	- Faulty DPS - Faulty logic resulted in higher DPS reading	- Replace DPS, or recalibrate if possible. - Conduct continuity loop testing and rectify logic

Table 9.1.29 Differential Pressure Transmitter

Equipment Type: Differential Pressure Transmitter

Preventive Maintenance

Maintenance Inspection	Acceptable Performance	Problem/Effect in case of low performance	Some possible causes in case of low performance - *SAMPLE*	Corrective/Preventive Action
Check DPT functionality with respect to control signal	Linear response to control signal on site/ True signal based on site condition	- Problem: Pump speed is not responding to DPT value. - Effect: Higher energy consumption	- Mechanical fault, fan operating on manual speed. - Lost signal from BMS - Faulty logic	- Return Pump to Auto mode - Check Pump speed response to DPT value and control logic on regular basis
Conduct continuity loop testing	Command/ Response received on frontend software and on-site equipment	- Problem: Command is not received on-site equipment - Effect: Higher energy consumption	- Mechanical fault, pump operating on manual speed. - Lost signal from BMS - Faulty logic	- Return pump to Auto mode - Check pump speed response to DPT value and control logic on regular basis
Compare differential pressure value on frontend and on-site	Command/ Response received on frontend software and on-site equipment	- Problem: Different DPT reading on frontend and on-site - Effect: Higher energy consumption	- Faulty DPT	- Replace DPT, or recalibrate if possible.

Predictive Maintenance

Maintenance Task	Acceptable Performance	Problem/Effect in case of low performance	Some possible causes in case of low performance - *SAMPLE*	Corrective/Preventive Action
Differential pressure value	Value within predefined limits	- Problem: DPT reading outside acceptable limits - Effect: Higher energy consumption	- Faulty DPT - Faulty logic resulted in higher DPT reading - Pump stuck on constant speed and not responding to DPT set point	- Replace DPT, or recalibrate if possible. - Conduct continuity loop testing and rectify logic - Check pump VFD and respond to control logic and rectify

Table 9.1.30 Flowrate/Velocity Meters

Equipment Type: Flowrate/ Velocity Meters

Preventive Maintenance

Maintenance Inspection	Acceptable Performance	Problem/ Effect in case of low performance	Some possible causes in case of low performance - *SAMPLE*	Corrective/ Preventive Action
Compare flow rate value on frontend and actual on-site value	98-100% match between actual site reading and frontend value	- Problem: Flowrate value on-site is higher/lower than the frontend. - Effect: if higher flow rate, higher energy consumption - Effect: if lower flow rate: the machine is not achieving intended function (i.e. AHU is no delivering required flowrate).	- Not calibrated flow meter. - Fault reading in BMS.	- Calibrate meter on a regular basis. - Check reading on a regular basis.
Conduct continuity loop testing	Command/ Response received on frontend software and on-site device	- Problem: Command is not received on-site equipment - Effect: possible higher energy consumption	- Lost signal from BMS - Damaged cables - Damaged BMS ports	- Check end to end BMS signal - Replace damaged cables - Replace damaged ports
Check flow rate response to software command (increase/ decrease)	Command/ Response received on frontend software and on-site device	- Problem: Motor(fan or pump) is not responding to BMS command - Effect: possible higher energy consumption, or lower delivered flowrate which is preventing the equipment of achieving its intended function	- Motor (fan/ pump) operating on manual - Motor is on Auto, but BMS signal is lost	- Operate motor on Auto - Check end to end BMS signal and rectify on a regular basis.

Predictive Maintenance

Maintenance Task	Acceptable Performance	Problem/ Effect in case of low performance	Some possible causes in case of low performance - *SAMPLE*	Corrective/ Preventive Action
Flowrate/ Velocity value	Value within predefined limits	- Site is higher/ lower than predefined limits. - Effect: if higher flow rate, higher energy consumption - Effect: if lower flow rate: the machine is not achieving intended function.	- Causes may be related to the equipment itself, BMS command, mechanical faults, etc.	- Case by case scenario, identify Corrective/ Preventive Actions and implement on regular basis

Table 9.1.31 Cooling Coil Temperature and Humidity Sensors

Equipment Type:	Cooling Coil Temperature and Humidity Sensors		
Preventive Maintenance			
Maintenance Inspection	Acceptable Performance	Problem/ Effect in case of low performance	Some possible causes in case of low performance - *SAMPLE*	Corrective/ Preventive Action
Check the sensor' ohmic value and actual temperature where it is installed	98-100% match between actual site reading and frontend value	- Problem: Actual temperature is lower/ higher than BMS value - Effect: If lower temperature, then higher energy consumption is recorded - Effect: if higher temperature, then insufficient cooling is supplied to the room	- Dirty sensor - Faulty Sensor	- Clean Sensor - Replace Sensor
Predictive Maintenance			
Maintenance Task	Acceptable Performance	Problem/ Effect in case of low performance	Some possible causes in case of low performance - *SAMPLE*	Corrective/ Preventive Action
Frontend recorded temperature value	Value within predefined limits	- Problem: Temperature is correct but lower/ higher than set limits - Effect: If lower temperature, then higher energy consumption is recorded - Effect: if higher temperature, then insufficient cooling is supplied to the room	- Two Way Valve is not responding to set point (refer to two-way valve section)	- Refer to two-way valve section

Table 9.1.32 Chilled Water Temperature Sensors

Equipment Type: Chilled Water Temperature Sensors

Preventive Maintenance

Maintenance Inspection	Acceptable Performance	Problem/ Effect in case of low performance	Some possible causes in case of low performance - *SAMPLE*	Corrective/ Preventive Action
Check the sensor' ohmic value and actual temperature where it is installed	98-100% match between actual site reading and frontend value	- Problem: Actual temperature is lower/ higher than BMS value - Effect: If lower temperature, then higher energy consumption is recorded - Effect: if higher temperature, then insufficient water temperature supplied to the equipment (i.e. AHU)	- Dirty sensor - Faulty Sensor	- Clean Sensor - Replace Sensor

Predictive Maintenance

Maintenance Task	Acceptable Performance	Problem/ Effect in case of low performance	Some possible causes in case of low performance - *SAMPLE*	Corrective/ Preventive Action
Frontend recorded temperature value	Value within predefined limits	- Problem: Temperature is correct but lower/ higher than set limits - Effect: If lower temperature, then higher energy consumption is recorded - Effect: if higher temperature, then insufficient water temperature supplied to the equipment (i.e. AHU)	- Issues with source of water (refer to chillers and heat exchangers sections)	- Refer to chillers and heat exchangers sections

Table 9.1.33 Space/Return Air Temperature and Humidity Sensors

Equipment Type: Space / Return Air Temperature and Humidity Sensors

Preventive Maintenance

Maintenance Inspection	Acceptable Performance	Problem/ Effect in case of low performance	Some possible causes in case of low performance - *SAMPLE*	Corrective/ Preventive Action
Check the sensor' ohmic value and actual temperature where it is installed	98-100% match between actual site reading and frontend value	- Problem: Actual space temperature is lower/ higher than BMS value or set point - Effect: If lower temperature, then higher energy consumption is recorded - Effect: if higher temperature, then insufficient cooling resulted in the space	- Faulty space temperature sensor - Wrong BMS reading	- Replace sensor - Check and correct BMS reading according to actual reading.
Check response of two-way valve to space/ return air temperature	Two-way valve open/close as per control logic	- Refer to two-way valve section	- Refer to two-way valve section	- Refer to two-way valve section

Predictive Maintenance

Maintenance Task	Acceptable Performance	Problem/ Effect in case of low performance	Some possible causes in case of low performance - *SAMPLE*	Corrective/ Preventive Action
Frontend recorded temperature value	Value within predefined limits	- Problem: temperature is lower/ higher than set limits - Effect: If lower temperature, then higher energy consumption is recorded - Effect: if higher temperature, then insufficient cooling resulted in the space	- Different possible causes could be in relation with mechanical equipment or control logic as described in previous sections	- Different possible corrective/ preventive actions related to mechanical equipment or control logic as described in previous sections

Table 9.1.34 Control Logic for all equipment controlled by BMS

Equipment Type: Control Logic for all equipment controlled by BMS				
		Preventive Maintenance		
Maintenance Inspection	Acceptable Performance	Problem/ Effect in case of low performance	Some possible causes in case of low performance - *SAMPLE*	Corrective/ Preventive Action
Control Logic for all equipment that is part of Energy Centered Maintenance process, for example: - AHUs - FCUs - FAHUs - Heat Recovery Units - Ecology units - Close Control Units - Pumps - Swimming pools cooling system - Chillers - Heat Exchangers - Fans - Lifts and Escalators - Travelators - Boilers - Cooling Towers - DX Units - MCC - VFDs	Control logic to be in accordance with system design sequence of operation	Case by case scenario, refer to previous sections	Case by case scenario, refer to previous sections	Case by case scenario, refer to previous sections

Chapter 10

Energy Centered Maintenance to Avoid Low Delta T Syndrome in Chilled Water Systems

This chapter is tackling another important topic that impacts the energy efficiency of equipment due to lack of proper maintenance and due to inefficient control logics, which is about the impact of low return chilled water temperature (so-called Delta T-syndrome) on chilled water system energy efficiency.

This chapter provides the basic definition of Low Delta T Syndrome and investigates some of the possible causes of it that result from lack of maintenance, and the ways of mitigating this problem.

This chapter describes what kind of maintenance tasks should be implemented during the planned preventive maintenance program to reduce or eliminate the causes of Low Delta T in chilled water systems. The maintenance tasks will be described in a similar manner as in previous chapters as part of energy centered maintenance model.

Effective maintenance measures and tasks can limit or reduce the impact of low Delta-T on equipment's energy consumption, system effectiveness and thermal comfort level in the buildings, thus saving energy and avoiding additional charges and costs.

LOW DELTA T SYNDROME DESCRIBED

Low Delta T Syndrome is almost a common problem in variable flow chilled water systems; Low Delta T is defined such that the actual difference between supply and return chilled water temperatures (Delta T) is less than the design value. In a simpler definition, low delta T oc-

159

curs when the return chilled water temperature from the building is lower than the design returns chilled water temperature, which is the case in most buildings operating with variable flow chilled water system.

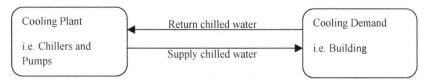

Figure 10.1 Schematic layout of supply/ return chilled water loop

Low delta T results in reducing the effectiveness of chillers and increases the energy consumption of the cooling plant. Lower return chilled water temperature decreases the temperature difference (ΔT) across the chiller as the water supply temperature is fixed to a setpoint; therefore the required chilled water flow rate will be larger to meet the expected cooling demand. The cooling plant performance is significantly reduced by lower operating chilled water Delta T than design conditions, and pumping energy are significantly increase to supply the increase in chilled water flow rate.

The following heat transfer equation represents the relation between the chilled water flow rate, chilled water temperature difference (Delta T) and chilled water flow rate:

$$Q = \dot{m} \; Cp \; \Delta T \hspace{3cm} (10\text{-}1)$$

Where:

    Q is the heat transfer, referred to as cooling load in kJ

    $\dot{m}$ is the water mass flow rate in m³/s.

    Cp Specific heat of water in kJ/kg.K

    ΔT is the temperature difference in Kelvin (K)

Equation 10-1 shows that the load is directly proportional to mass flow rate and temperature reference. Therefore, if the delta T remains constant, the required mass flow rate will remain constant to give the same load, while if the temperature difference drops, the mass flow rate should increase to meet the same load.

The same principle is happening in the chilled water system that is suffering from low Delta T Syndrome, where the system fails to maintain a constant delta T.

In this case, to keep the same heat transfer (cooling load generated by the chillers), the mass flow (chilled water flow rate) should increase in case the Delta T (temperature difference between supply and return chilled water temperatures) drops. The increase in chilled water flow rate will increase the pumping energy of chilled water pumps, and it will reduce the chillers effectiveness which will result in adding more chillers into operation to meet the required chilled water flow rate and cooling demand of the building.

Chillers effectiveness is calculated based the ratio between actual Delta T value to design Delta T value as shown in the following formula:

Chiller Effectiveness = ARCWT – ASCWTDRCWT – DSCWT x 100%   (10-2)

- Actual return chilled water temperature (°C), referred to as ARCWT.
- Actual supply chilled water temperature (°C), referred to as AS-CWT.
- Design return chilled water temperature (°C), referred to as DRCWT.
- Design supply chilled water temperature (°C), referred to as DSCWT.

If the actual chilled water temperature drops by a certain value from the design chilled water temperature, it will have a direct impact on the chillers effectiveness. For example, if the required delta T according to design is 8.0 °C, while the actual delta T is 6.4 °C, the chiller effectiveness drops from 100% to 80%. In such a case the same existing chillers cannot meet the building demand because the effectiveness has dropped, which will require additional chillers to be put into operation to meet the required demand. It is important to highlight that the actual number of chillers needed to operate at part load operation is decided based on chillers capacity and based on the control logic of the cooling plant.

Low delta T has a direct impact on energy consumption of the facility and effectiveness of the chilled water system. As explained earlier, when the Delta T drops, it increases the energy consumption of the chilled water pumps and chillers by increasing the pumping energy and by turning on more chillers and pumps into operation to meet the same cooling demand.

Low Delta T Syndrome results from the inefficient use of chilled water amount at the building and plant side, as well as due to lack of maintenance of chilled water system components which is yielding a lower return chilled water temperature than the design value. Maintaining the right delta T in chilled water systems increase the energy efficiency of the cooling plant and reduces potential energy waste during plant operation.

MAINTENANCE RELATIONSHIP

Lower Delta T Syndrome results from the inefficient use of chilled water at the building and plant side yielding a lower chilled water return temperature than the design value. There are multiple causes of Low Delta T which could be related to design issues, installation issues, and most importantly due to improper operation and maintenance.

Energy centered maintenance model focuses on the issues appear during the operation of the equipment and provides guidance on identifying the root-cause of the problem and the way to mitigate by conducting certain corrective and preventive actions.

This section identifies some of the possible causes of low delta T that are related to lack of maintenance. However, multiple types of research and studies have been conducted by other researchers who identified the possible causes of the problem from the design perspective, some of those causes will also be listed for the benefit of the reader.

Steven T. Taylor has issued a paper titled "Degrading Chilled Water Plant Delta-T: Causes and Mitigations" in which he listed most possible causes of the problem and the way to mitigate it. Another paper by Donald P. Fiorino titled "Achieving High Chilled Water Delta Ts" also discussing 25 best practices to achieve high chilled water temperature difference. Some of the points considered by both authors are discussed in this chapter.

CAUSES CAN BE AVOIDED DURING DESIGN STAGE

Some of low Delta causes can be due to either design faults related to the selection of an improper type of equipment, or poor control logic of chilled water system components. Those causes can be eliminated by taking into consideration the right factors in new designs which can mitigate this problem:

1. **Use of Constant Flow Chilled Water System:**
    The use of constant flow chilled water system maintains the same supply chilled water flow rate amount to Fan Coil Units (FCUs) until the set point temperature inside the room is achieved. Till then the chilled water temperature varies with the load, and it could drop below design return temperature and causes low delta T.

Table 10.1 Use of Constant Flow Chilled Water System

Cause	Corrective/ Preventive Action
Use of constant flow chilled water system	Use variable flow chilled water system instead of constant flow along with selecting proper equipment size, control valve type and size and control logic.

2. **Use of 3 Way control valve in variable flow chilled water system:**
    Most control valves used in variable flow chilled water system are 2-way control valves. However, designers tend to keep few 3-way valves in the system to allow chilled water flow rate to circulate. 3-way valves are not connected to the load (not to a Fan Coil Unit). Instead, they may be provided at the end of the chilled water network for water circulation. In this case the same supply chilled water temperature returns to the cooling plant, which affects the total return chilled water temperature from the building and could lead to low delta T.

Table 10.2 Use of 3 Way control valve in variable flow chilled water system

Cause	Corrective/ Preventive Action
Use of 3-way control valve in variable flow chilled water system	Don't use of 3-way valves in variable flow systems for chilled water circulating. The intent of circulating the water is to reduce the time the chilled water will reach to an FCU when it requires it. The reality is that the chilled water flow is always circulating in the system even in low demand time.

Table 10.3 Cooling Coil Selection for Low Delta T than Design

Cause	Corrective/ Preventive Action
Cooling Coil Selection for Low Delta T than Design	Always ensure that the cooling coil is selected for the same design delta T temperature before manufacturing the units and installing it at the site.

3. **Cooling Coil Selection for Low Delta T than Design**
    Sometimes fan coil units cooling coils are selected for lower sup-ply/return temperature difference than design. In this case, and even if the units are functioning its best performance, the return chilled water temperature will always be less than the required temperature just because the coil is designed for lower temperature difference and it will cause a low delta T.

4. **Oversized Airside Equipment**
    Oversized Fan coil units and air handling units receive more chilled water flow rate than the anticipated cooling load in the room it's serving, which leads to incomplete heat transfer between the chilled water and the room air that results in returning chilled water on a lower temperature than design.

Table 10.4 Oversized Airside Equipment

Cause	Corrective/ Preventive Action
Oversized airside equipment	Always ensure that airside equipment is selected according to cooling load demand

5. **Chilled Water Pumps selected with higher pump head than actual**
    If chilled water pumps are selected with higher pump head, the pumps will tend to pump more chilled water flow rate to the build-ing during high load demand, the flow rate, in this case, will be more than what is required and will lead to return the water with lower temperatures that design return temperature.

Table 10.5 Chilled Water Pumps selected with higher pump head that actual

Cause	Corrective/ Preventive Action
Chilled Water Pumps selected with higher pump head that actual	Pumps should be selected on the right head, and index location should be determined in the chilled water circuit to ensure the right amount of chilled water flow rate is supplied to the building at all time.

6. **Use of Pressure-dependent Control Valves**

    Pressure-dependent control valves are usually selected based on system pressure, which leads in most cases to valves being oversized. Oversizing allows more chilled water flow rate to circulate through the airside equipment according to system pressure; which results in lower return water temperature.

Table 10.6 Use of Pressure-dependent Control Valves

Cause	Corrective/ Preventive Action
Use of Pressure Dependent Control Valve	Use of Pressure Independent Control Valves provides a steady flow for the airside equipment regardless of pressure fluctuation in the system. Which always ensure minimum flow is supplied to the airside equipment, and optimum delta t is achieved.

CAUSES CAN BE AVOIDED
DURING OPERATION AND MAINTENANCE

Low delta T causes discussed in this section are a result of poor operation and maintenance. Those causes may be the root of the problem or may contributed to it. Therefore, the maintenance personnel should conduct a detailed analysis to ensure that the root causes are known, and the associated corrective and preventive actions are put in place to ensure the issue will not happen again.

Preventive actions should be then implemented on a regular basis as maintenance inspections and tasks; those tasks should be defined in the planned maintenance job plans proactively on a predefined frequency. Maintenance frequency should be determined by the process mentioned in Chapter 6 in a way that prevents the problem from happening again.

The cost effectiveness of the process should be calculated based on all costs associated with the maintenance job plan compared to the energy saved and the waived low delta T charges imposed by cooling provider (if any)

1. **Chilled Water Control Valve Left in Open Position:**
    In some cases, it was found that the two-way chilled water control valve is left open position manually or stuck on open position. Which prevents the control valve from doing its function in allowing the required chilled water flow rate from passing through the airside equipment, which causes overflowing of chilled water flow at part load conditions. In this case, the heat transfer between the air inside the room and the chilled water flowing inside the cooling coil is not complete which leads to low return chilled water temperature.

Table 10.7 Chilled Water Control Valve Left in Open Position

Cause	Corrective/ Preventive Action	Inspection Frequency
Chilled water control valve left on open position	Check chilled water control valve position and ensure it is functioning on Auto position and modulating chilled water flow according to demand. Ensure control valve is not stuck on open position	Monthly

2. **Chilled Water Control Valve is Not Responding to Space Temperature:**
    In some cases, the two-way control valve doesn't modulate (close/open) by space temperature. If the set point temperature inside the room is 24.0 °C and the actual room temperature is also 24.0 °C then

the control valve actuator should close the two-way and prevent chilled water from flowing through the cooling coil.

If the control valve remained open the chilled water flow would circulate inside the cooling coil with minimal heat transfer with the room air; the heat transfer will not be sufficient to warm the return chilled water temperature. Hence water will return to the cooling plant lower than required and contributes to low delta T.

Table 10.8
Chilled Water Control Valve is Not Responding to Space Temperature:

Cause	Corrective/ Preventive Action	Inspection Frequency
Chilled water control valve is not responding to space temperature	Check chilled water control valve position and ensure it is functioning on Auto position and modulating chilled water flow according to demand. Ensure control valve is not stuck on open position	Monthly

3. **Dirty/ Clogged Cooling Coil or Air Filter**

    If the cooling coil or the air filters of the airside equipment are clogged, the room air will not efficiently pass through the cooling coil. And the amount of the air that will pass through the coil will not be sufficient to complete the heat transfer between with the chilled water passing inside the cooling coil, which will lead to low return chilled water temperature.

    At the same time, since the airflow rate is lower than the commissioned airflow rate, the space set point temperature will not be achieved. Which will maintain the control valve on open position trying to achieve the required room temperature, which will also allow the chilled water to pass through the cooling coil without sufficient heat transfer and the return chilled water temperature remains lower than design.

10.9 Dirty/ Clogged Cooling Coil or Air Filter

Cause	Corrective/ Preventive Action	Inspection Frequency
Dirty/ clogged cooling coil or air filter	Check and clean cooling coil and air filters on a regular basis.	Cleaning frequency differs between circulating airside equipment and fresh air units. Fresh air units require a higher frequency of cleaning which could be bi-weekly or monthly.

4. **Opened Bypass Lines on Airside Equipment**

    During testing and commissioning stage, the bypass line is usually provided for flushing chilled water pipe network, which should be closed after flushing is done. In some cases, those bypass lines are left open which results in flowing the supply chilled water directly to the return lines with the same supply chilled water temperature. This is resulting in a high impact on return chilled water temperature and results in low delta T.

    Table 10.10 Opened Bypass Lines on Airside Equipment

Cause	Corrective/ Preventive Action	Inspection Frequency
Opened bypass lines on air side equipment	Check and ensure bypass lines are closed after flushing	Annual, or whenever system flushing takes place

Chapter 11

Energy Centered Maintenance in Data Centers

ECM TERMINOLOGY AND CHARACTERISTICS

Data centers usually operate 24-7, 365 days a year. Need troubleshooting and monitoring all the time by both operations and maintenance. Keeping the servers running under the right environment without getting excessive heat or excessive water is the primary goal. They are declared by the organization as a mission critical facility. The data center can be in a facility by itself or in a facility containing other functions or organizations. It can be gigantic (multi-rooms) or located in one small room.

Data centers have HVAC, chillers, cooling towers, CRACs (computer room air conditioning), backup power (uninterruptable power supply-ups), boilers, generators, lights, servers, and other equipment.

Power to the facility remains flat through the year (no seasonal changes as seen in other facilities) and costs around or more than $50,000 a month depending on size and equipment inside. Data centers use around 2% of energy consumed in the U.S. yearly.

The key definitions that apply to a data center are:

CRAC—Computer Room Air Condition
   PDU—Power distribution units—Distributes power to servers, performs power filtering and load balancing and provides remote monitoring and control.
   Plenum—Concealed area that enables hot air flow back to CRAC.
   PRO DC—A software program developed by DOE to identify energy savings in a data center.
   PUE—Power use effectiveness—total facility power/IT equipment & operations use in kW.
   Rack—Container that holds the servers.

Raised Floor: Enables the cool air or cold water cables to cool the servers.

Server: a computer that computes or provides service to a network or internet.

UPS: Uninterruptable power supply-provides backup power in case of an interruption to main power supply.

The data center uses a lot of energy. ECM is a natural tool to use to decrease energy consumption while ensuring data center operations are productive and safe from the dangerous items that can quickly bring a data center to a halt.

IF YOUR DATA CENTER IS COLD,
THEN THESE SOLUTIONS WILL MAKE YOU GOLD

If the data center is cold, then ECM is needed immediately to re-duce energy consumption from 30 to 40%. Your approach is six actions.

Action 1. Place the servers in a hot/cold configuration with only cold air on one side and hot air coming from the server back on the other side.

Action 2. Check the temperatures in the server room. It is prob-ably around 66 degrees Fahrenheit. The ASHRAE Standard allows significantly higher temperatures now than in the past. By raising the temperatures, less air conditioning is needed, less work on the cooling towers, and you may be able to retire one or more CRACs since they may not be needed anymore. Many data centers are too cold and dry. In 2008 ASHRAE adjusted its guidance around supply air temperatures to 64.4°F at the lowest extreme and 80.6°F at the high end (up from its pre-vious recommended high of 77°F). Many data centers have been setting temperatures much lower, as low as 55°F. By raising the temperature even one degree, data centers can achieve energy savings. For humidity, ASHRAE has also loosened its guidelines, increasing the high end to 60% relative humidity (up from 55% in 2004).

Action 3. Separate the cold air from mixing with the hot air on the way back to the CRAC. You may need a contractor to plan and have this built and put in place. Panels are used including panel doors to seal off the cold aisles. This project will have less than two years payback. Be sure to include the Fire Marshal and the Maintenance Department or Contractor in on the beginning of the planning and execution.

Action 4. Using a wattmeter, find out for each server the watts being used to power the server and compare to the nameplate what should be utilized. The old servers could easily be "energy hogs" and if so, should be replaced.

Action 5. Measure the utilization of each server. If the utilization is low, then have the programmers increase the use. This action will provide more server efficiency, thus reducing some servers for being used and lower the energy consumed. There will need to be some servers with lower utilization to be used as a backup to the main ones.

Action 6. Calculate either with actual data or estimated data, the P.U.E. (power usage effectiveness). Power usage effectiveness—Total facility power (kW)/IT equipment power (kW). A PUE of 1.6 to 1.3 is good, with 1 being the goal. The lower the number, the larger percentage of the total power furnished to the facility is used by the data center operations. (80% of the power furnished used by the IT equipment would be excellent. This is the DCIE indicator whose reciprocal is the PUE. $1/.8=1.25$)

- Facebook's new data center in Forest City, NC, has a PUE between 1.06 to 1.08.
- A high PUE should drive improvement in your data center.

Taking these six actions will lead to gold or at least a lot of energy and money saved.

Chapter 12

Measures of Equipment and Maintenance Efficiency and Effectiveness

Lead (Key Performance Indicators) and Lag (Key Result Indicators)

Lagging indicators are normally "output" or "outcome" oriented, *easy to measure* but *difficult to improve* while leading indicators are normally "input" oriented, *challenging to measure* and *easy to influence*. Lead indicators are measures that **"drive"** the **performance** of lag ones. Normally, lead indicators measure processes and activities. Lag Indicators are outcomes of these processes. The lead indicators are often called key performance indicators (KPIs).

KRIs tell management and stakeholders if the organization is operating well. They do not provide any information or insight into what activities or actions were successful or not. They do determine whether the organization's end goals were met or not.

While KRIs cannot help you achieve or improve upon the organization's goals, KPIs focus on the actions that lead to the results. KPIs provide the information that's important in creating strategies and aligning goals.

Key results indicators(KRI) measure the *results of goals and business strategies*. Key performance indicators (KPIs) measure actions that are important and influences or impacts the **KRIs**. Most organizations need a mixture of KPIs and *KRIs*. *Whether* an organization meets their corporate goals, most times depends on if there are helpful KPIs to influence the KRIs outcomes.

MAINTENANCE GROUP INDICATORS

The maintenance group or department should have a combination of KPIs and KRIs. The KRIs are results and normally are traditional in-

Figure 12.1 Introduction to Lead and Lagging Indicators

Figure 12.2 Lead and Lag Indicators Relationships

Figure 12.3 KPIs(Lead) Drives KRIs(Lag)

dicators. The KPIs drive the KPIs and are drivers and as such are very actionable. Once they are completed, they are no longer actionable and should be discontinued. For example, percent people committed. Once everyone is committed, we no longer need to measure and display this KPI. We now need to find another, if possible, that will drive the particular KRI.

The maintenance department or group should have measures they are in control of and are actionable only if by having a KPI to influence the KRIs favorably. It is so easy to select performance indicators that provide information but are not actionable that can help improve the maintenance operations or equipment reliability. Some traditional maintenance indicators such as mean time between failure (MTBF) is impacted by others such as affected by the ordering and receiving of the original equipment, the manufacture's quality of production, the equipment design, maintenance tasks performed or not performed, and quality of the parts used and even workmanship.

The planning group should already have some performance indicators before implementing energy centered maintenance (ECM). Some of these could be impacted by ECM and be impossible to distinguish if the preventative maintenance or predictive maintenance achieved the results or ECM. Therefore, the ECM performance indicators selected should be controlled in the ECM Process and be actionable to the ECM goals and objectives.

Setting goals are important for any maintenance group to improve their efficiency, productivity, and effectiveness. The ECM technicians or mechanics will have to abide by the rules and be a part of the operations helping achieve these goals. What are these goals? Of course, these goals will vary by organization, but several will probably be included by most organizations. They are:

- Goal 1: Up time greater than 90%
- Goal 2: Percent emergency work less or equal to 10%
- Goal 3: Training days per technician or mechanic is 6-9 Days a Year
- Goal 4: Schedule achievement is greater than 90%
- Goal 5: Technicians or mechanics to a maintenance planner 25 to 1.
- Goal 6: Percent overtime less than 8%
- Goal 7. Maintenance cost to estimated equipment replacement cost is 2.5% or less

The technicians or mechanics performing ECM tasks will have some impact, although not the primary purpose, on Goal 1. They will adhere to Goal 3, 6 and 7, help achieve Goal 4 and be included in Goal 5. At least four performance measures jump out from these goals. Energy centered maintenance indicator #1 is schedule achievement. The indicator target is 90% or higher. The second indicator, ECM indicator #2 is% overtime. Since the ECM tasks are not fixing breakdowns, overtime should rarely be used. There will be occasions where overtime may be necessary to inspect a machine or repair it on off peak hours. However, the target should be 1% or less, considerably lower than 8% for other maintenance. ECM performance indicator #3 is ECM cost divided by the estimated en¥ergy savings from doing the inspections, maintenance, repair or replacement. This ladder indicator is an ECM payback indicator with a target of 3 years or less. ECM indicator # 4 is% of technicians or mechanics trained in ECM. The target is 100% in first six months.

The work done by maintenance needs to support the business aims and operating strategy. The ideal way to show that is to have maintenance performance linked to the reasons your company is in business. This is harder for most other maintenance types to do than for ECM in that every organization needs the energy to perform its mission providing its products or services. The energy costs can be substantial, sometimes up to 30% of total operational costs. Therefore, the energy cost reduction (kWh Reduced x Cents per kWh = Cost Savings) is a welcomed initiative by any company, university or business. Performance Indicator #5 & #6 is energy consumption reduced and energy costs savings respectively. Along with these, the key result indicator, electricity or energy intensity, becomes our ECM key results indicator #7. The target is 5% reduction during the first year. It is defined for electricity as kWh consumed / total gross square feet of facilities maintained.

Effective maintenance is defined as doing what is right. Training of technicians or mechanics and having detailed precise maintenance instructions coupled with work order planning and scheduling will favorably impact the maintenance effectiveness. Doing efficient maintenance is doing the right work required that improves reliability or accomplishes its purpose with the least resources and with productivity. Successful maintenance is not fixing things but not having to fix them. Therefore, not having to do the maintenance would be an excellent indicator showing the maintenance effectiveness. This is a useful performance indicator but tough to measure. It is important that when we select a group of maintenance key performance indicators we pick the ones that let us improve both equipment reliability, maintenance performance, workforce productivity, goals achievement and not just tell us the problems experienced in our businesses or organizations.

ECM can contribute to equipment reliability improvement, even though that is not its primary objective or purpose. Some organizations may want a Key Result Indicator to measure if equipment reliability is being improved or not in the long term. For example, some equipment inspected in ECM that were targeted for possible energy excessive use could measure mean time between failure for these. The industry mean time between failure could be obtained.(ECM indicator # 8 MTBF) The target is three years plus. Then, measuring when they have an actual failure could show whether equipment reliability has been improved or not. For example, the equipment to be measured could be:

Table 12.1 Targeted Equipment with Present MTBF

Equipment	MTBF (Not Accurate)
Pumps	3 Years
Compressors	4 Years
Air Handling Units	3 Years
Boilers	2.5 Years
Cooling Towers	2.5 years
Elevators	2 Years
Water Supply System	3 Years

KRI ECM Indicator #9 is Per Cent Improvement in five years (could be any year after 4) can be calculated for individual equipment or as an aggravate. The target is 10% increase. This is an optional Key Result Indicator.

For measuring maintenance, the areas that often are measured are:
1. The Efficiency of the Maintenance Group.
2. The Effectiveness of the Maintenance Group.
3. The Productivity of the Maintenance Group.
4. The Performance of the Maintenance Group.
5. Progress Against Maintenance Goals or Adherence to Maintenance Policies.
6. Plant Reliability Goals
7. Equipment Availability along with Down Time.
8. Schedule Achievement
9. Business Value of Maintenance
10. The Quality of Maintenance.
11. Overall Equipment Efficiency

OVERALL EQUIPMENT EFFICIENCY (OEE)

KRI #10 is OEE = Availability x Performance x Quality
http://www.oee.com/calculating-oee.html

Let's say the target is 87%. These three areas are usually measured separately and then multiplied to give the overall equipment efficiency. Therefore, individual targets can be established for each of the three factors and when they are multiplied to get OEE would equal 87% (providing the target is achieved).

The reasons their low OEE are poor performance of the machines due to extensive set-up and cleaning periods, production downtime, inefficient processes, and poor planning.

These problem areas can be removed by improving the transparency of machine utilization times, collecting and analyzing potential sources of loss, and always providing information on production processes in real time.

KRI # 11 is Machine Efficiency

Machine efficiency is

Efficiency = Output value / Input value

For example, if a machine needs 10 KW to run and produces 8 KW, its power efficiency is $8/10 = 0.8$ or 80%.

Efficiency is always between 0 and 1 (or 0 and 100 if expressed as a percentage.)

ECM KRI # 12 is OEE Availability

Availability includes all events or happenings that stop planned production sufficient enough where it makes sense to track a reason for being down (several minutes).

Availability is defined as the ratio of run time to planned production time where planned production time is the total time that equipment is expected to produce.

Availability takes into account down time loss, and is calculated as:

Availability = Operating time / planned production time

Down time can be a result of lack of maintenance, ECM or preventative or predictive.

ECM KRI 13 is OEE Performance

The performance takes into account speed loss. It's formula is:

Performance = the ideal cycle time / (operating time / total pieces)

Ideal cycle time is the minimum cycle time that your process can be expected to achieve in optimal circumstances. It is sometimes called *design cycle time, theoretical cycle time* or *nameplate capacity*.

Since run rate is the reciprocal of cycle time, *performance* can also be determined as:

The performance = (total pieces/operating time)/ideal run rate.

ECM KRI # 14 is Quality

Quality takes into account quality loss, and is calculated as:

Quality = good pieces/total pieces

OEE scores provide very useful information. It is an accurate picture of how effectively your manufacturing process is running. In addition, it enables improvements to be tracked over time.

The OEE score by itself doesn't provide us any insights as to the underlying causes of lost productivity. This is the role of availability, performance, and quality.

In the preferred calculation you get the best of both worlds. A single number that captures how well you are doing (OEE) and three numbers that capture the fundamental nature of your losses (availability, performance, and quality).

Here is an interesting example. Look at the following OEE data for two sequential weeks.

Planned Production Time

The OEE calculation begins with planned production time. So first, exclude any shift time where there is no intention of running production

Table 12.2 Overall Equipment Efficiency for Two Sequential Weeks

OEE Factor	Week 1	Week 2
OEE	77.1	78.1%
Availability	88.0%	88.1%
Performance	89.4%	89.81%
Quality	98.0%	97.92%

Item	Data
Shift Length	8 hours (480 minutes)
Breaks	(2) 15 minute and (1) 30 minute
Down Time	43 minutes
Ideal Cycle Time	1.0 seconds
Total Count	20,100 do dads
Reject Count	418 do dads

(typically breaks).

Formula: Shift length – breaks

Example: 480 minutes – 60 minutes = 420 minutes

Run time

The next step is to calculate the amount of time that production was actually running (was not stopped). Remember that stop time should include both unplanned stops (e.g., breakdowns) and planned stops (e.g., changeovers). Both provide opportunities for improvement.

Formula: Planned production time – stop time

Example: 420 minutes – 43 minutes = 377 minutes

Good Count

If you do not directly track good count, it also needs to be calculated.

Formula: Total count – reject count

Example: 20,100 do dads – 418 do dads = 19,682 do dads

Availability

Availability accounts for when process is not running (both unplanned stops and planned stops).

Formula: run time/planned production time

Example: 373 minutes/420 minutes = 0.8881 (88.81%)

Performance

Performance accounts for when the process is running slower than its theoretical top speed (both small stops and slow cycles).

Formula: (ideal cycle time × total count)/run time

Example: (1.0 seconds × 20,100 do dads)/(373 minutes × 60 seconds) = 20100/22380=0.8981 (89.81%)

Quality

Quality accounts for manufactured parts that do not meet quality standards.

Formula: good count/total count

Example: 19682/20100 do dads= 0.9792 (97.92%)

OEE

Finally, OEE is calculated by multiplying: Availability × performance × quality

Example: 0.8881 × 0.8981 × 0.9792 = 0.78.1 (78.1%)

OEE can also be calculated using the simple calculation.

Formula: (good count × ideal cycle time)/planned production time

Example: (19,682 widgets × 1.0 seconds)/(420 minutes × 60 seconds) = 19,682/25,200=.781 (78.1%)

The result is the same in both cases. The OEE for this shift is 78.1%.

ECM INSPECTION

The schedule achievement was measured in KRI # 1. We also need to know the percentage of equipment measured on time (KPI # 15) and when equipment is using excessive energy, How many, or percent were fixed in 1 day or less (KPI #16).

INDICATOR CHECKED

We know the reasons indicators are developed. We can use a table or matrix and check to see if we adequately covered the important areas, and see if we have any serious gaps in our measurement coverage. See Table 12.3.

It is easy to get a flood of indicators. Be sure only to select those that are meaningful and develop a data collection sheet for each.

Figure 12.4 Data Collection Sheet
1. Indicator Title:
2. Data Source:
3. Frequency:
4. Formula:
5. Graph Type:
6. Description of Data:
7. Person Responsible:

Most of the areas are self-explanatory except description of data. For example, if data are included for weekends or not included, people would know by this description. Each indicator should have a data collection sheet.

Table 12.3 Indicators Check

Reasons for Indicators	KPI (Lead)	KRI (Lag)	Coverage Adequate Yes or No?	Possible new Indicator or One needed
1. The Efficiency of the Maintenance Group	4. % Trained in ECM, 16. % Equipment Using Excessive Energy Fixed in 1 Day or less	11. Machine Efficiency	Yes for ECM, No for Other Maintenance	% Tools Requested and Filled (KPI)
2. The Effectiveness	15. % Equipment Inspected on Time, 16% Equipment Using Excessive Energy Fixed in 1 Day or less	5. Energy Consumption, 6. Energy Cost, 7. Energy Intensity, 8. MTBF, 9. % Improvement in MTBF, 12. Availability & Down Time	Yes	
3. The Productivity			Not Needed for ECM	
4. The Performance	15. % Equipment Inspected on Time, 16. % Equipment Using Excessive Energy Fixed in 1 Day or less	10. OEE	Yes	
5. Goal Achievement Progress or Adherence to Policy	15. % Equipment Inspected on Time, 16. % Equipment Using Excessive Energy Fixed in 1 Day or less	2. Over Time	Yes	Need a Policy & Procedure
6. Plant Reliability	15. % Equipment Inspected on Time, % Equipment Using Excessive Energy Fixed in 1 Day or less	8. MTBF, 9. % Improvement in MTBF	Yes for ECM, No for Other Maintenance	Not ECM's Main Purpose

Table 12.3 (*Cont'd*) Indicators Check Continued

Reasons for Indicators	KPI (Lead)	KRI (Lag)	Coverage Adequate Yes or No?	Possible new Indicator or One needed
7. Equipment Availability	15. % Equipment Inspected on Time, 16. % Equipment Using Excessive Energy Fixed in 1 Day or less	12. Down Time & Availability	Yes	
8. Schedule Achievement	15. % Equipment Inspected on Time, 16. % Equipment Using Excessive Energy Fixed in 1 Day or less	1. Schedule Achievement, 11. Machine Efficiency13 OEE Performance	Yes	
9. Business Value of Maintenance		3. Payback, 5. Energy Consumption, 6. Energy Cost, & 7. Energy Intensity	Yes	
10. Quality of Maintenance		14. Quality	Yes	
11. Overall Equipment Efficiency	15. % Equipment Inspected on Time, 16. % Equipment Using Excessive Energy Fixed in 1 Day or less	10. OEE	Yes	

TARGET SETTING

Setting a target is not about magically pulling a figure out of the air. To set a target, you must first know where you are, what you want to achieve, and then be able to determine realistically, but stretch estimates that represent challenging amounts of improvement needed to attain the target.

How do we define a target? A target and stretch target can be defined as:

Targets: the desired level of performance you want to achieve on your indicator after a period that shows success from the countermeasures you have implemented along the way.

Stretch Target: A challenging but an attainable target that can be reached with reasonable but slightly accelerated effort.

The steps in setting a target are:

1.   Define what level or present status.
2.   Determine "how much" or what you want to achieve and "by when."
3.   Establish the timeframes you need the target to be attained.

In setting the targets, it is good to set a SMART one by using and adhering to the SMART characteristics:

Specific: What you plan to achieve is understood.

Measurable: There is an indicator or measure that shows whether you have achieved it or not.

Achievable: With the resources available.

Realistic: Stretch, but possible.

Timeframe: From and to are specified.

Targets drive performance. Therefore, set good ones but make them achievable. Targets and indicators go together. With every good indicator, there is always a good target.

Chapter 13

Energy Savings Verification

BASELINE

The typical way energy savings is achieved is to implement an action, a countermeasure or a project and when they checked, the targeted kWh reduces and is verified as a savings or reduction when verified against the baseline.

Having a baseline is critical. It gives you something to compare current with to see if any improvement has been made or not.

Energy baseline is the basis for an initial energy assessment of the building, it summarizes the building's current energy performance and helps the energy management team to draw basis of where the complete energy management processes should start. It will also assist the energy management team to understand how energy expenditure contributes to operating cost.

To develop a baseline, obtain three years of electricity bills and plot the monthly consumption on a line graph. Energy baseline can be developed and evaluated by energy management team based on energy data history and real-time metering. An energy baseline is an approach in which all energy inputs to a facility are accounted for; it is a preliminary energy assessment for the current energy behavior of the building before conducting detailed energy analysis. Energy baseline is the analysis of the energy use and consumption based on measurement and other data obtained from metering and energy bills.

The information extracted from the preliminary energy baseline assessment will allow the energy management team to benchmark the building energy performance with other buildings, establishing energy performance indicators and setting energy management targets.

Energy baseline will help the energy management team to:
• Identification of the current energy consumption of the facility.
• establishing internal energy consumption benchmark.
• benchmarking energy consumption with other similar facilities.

- identification of all internal and external factors affecting the energy consumption of the facility.
- identification of significant energy consumption equipment.
- identification of energy flows in the facility.
- estimate future energy use and consumption
- identify, prioritize and highlight opportunities for improvement.

Energy baseline puts the basis to ask the how we can reduce electricity consumption? We can reduce either the connected load or the operating hours. Finding motors running 24 hours a day when they were only required to run 7 hours will reduce the operating load for motors. In universities and colleges, during holidays and times when rooms are not occupied, the equipment in the room, the air conditioning or heating and ventilation can be turned off saving energy consumption. Checking each item necessary operating time and if smaller reducing the operator time. Changing operation hours for some equipment from peak hours to off-peak can save considerable energy costs. When replacing any equipment, it may be possible to reduce the connected load by purchasing a more energy efficient one such as Energy Star and replacing the item when it uses excessive energy or about to break down.

Example of Energy Baseline

To develop a baseline, obtain three years of electricity bills and plot the monthly consumption on a line graph.

Table 13.1 kWh By Month and Year

Yr/Mo	Jan	Feb	Mar	Apr	May	June
2013	42090	44090	39800	38678	38900	41267
2014	43000	43987	40122	38980	39100	40900
2015	39800	40378	40000	36789	37500	39870
	July	Aug	Sept	Oct	Nov	Dec
2013	43300	44500	46780	45189	46780	47990
2014	44600	44190	46000	44236	46700	46789
2015	41380	44200	44560	42900	44360	44350

Notice the seasonality of the yearly graphs. Years 2013 & 2014 have almost same trends. The year 2015 shows some reduction of kWh consumption. In late 2014 some energy projects were implemented and resulting reduction is happening. The percent improvement is 518595 − 496087 = 22508/518595 = .043 = 4.3%.

Figure 13.1 kWh Monthly Consumption By Year

Table 13.2 Year Totals

2013	2014	2015
519364	518595	496087

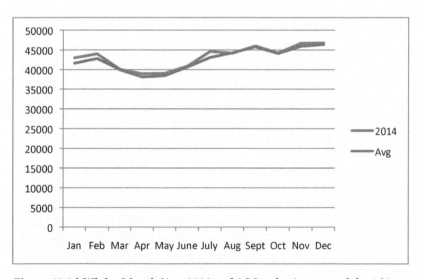

Figure 13.2 kWh by Month Year 2014 and 3 Months Average of the 3 Years

Therefore, either 2013 or 2014 would be a better baseline. However, let's see how the yearly monthly averages compare to 2014.

The year 2014 and the monthly averages trend are almost identical. Therefore, the year 2014 is selected as the baseline that all possible improvements in the future will be compared with for verification.

How can we reduce electricity consumption? We can reduce either the connected load or the operating hours. Finding motors running 24 hours a day when they were only required to run 7 hours will reduce the operating load for motors. In universities and colleges, during holidays and times when rooms are not occupied, the equipment in the room, the air conditioning or heating and ventilation can be turned off saving energy consumption. Check each item necessary operating time and if smaller reducing the operator time. Changing operation hours for some equipment from peak hours to off-peak can save considerable energy costs. When replacing any equipment, it may be possible to reduce the connected load by purchasing a more energy efficient one such as Energy Star and replacing the item when it uses excessive energy or about to break down.

ENERGY BASELINE, ENERGY TARGETS,
AND ENERGY PERFORMANCE INDICATORS

Developing energy baseline alone does not enable the energy management team to determine if energy is being used efficiently on continues basis or not. Developing energy baseline should be part of the operating culture of the facility, and should be developed on a regular basis to ensure that energy is consumed in an efficient way on a continues basis. Therefore, energy baseline should be transformed to monthly energy targets and energy performance indicators (EnPIs) which should be developed for the building by the energy management team.

Setting energy targets is important because it provides a direction of what needs to be achieved within a certain time frame. Therefore, after energy benchmarking is set, and energy performance indicators are defined and analyzed, the energy management team shall be able to set energy targets to improve performance. Energy targets must be measurable and achievable depending on information from benchmarking.

Energy targets shall identify target levels including what savings are included, how savings are to be evaluated, and specific energy performance indicators and baselines to be used. The moment the energy saving targets are set, the effect of energy management process shall start by conducting the energy audit. Further, where energy targets found to be achievable via systems retrofit, the energy management team shall develop a business case feasibility study comparing the investment cost with the return on investment value over a specified period to confirm its feasibility.

The ability to measure the real performance is the foundation for improvement and optimization.

Energy management key performance indicators are used to track the percentage of organization's goals for energy and carbon reduction.

Three performance indicators are critical for managing energy consumption: energy utilization index (EUI), energy cost index (ECI) & energy productivity index (EPI). These indicators provide a measure of which the degree of performance can be related to a given target.

Energy utilization index (EUI): Ratio of the energy consumed in a building over a given period (year), per gross square meter of conditioned space.

Table 13.3 Energy Utilization Index (EUI)

Value Driver	Unit	Measurement	Energy Reduction Target
Energy Utilization Index (EUI)	KWH/m²/Year	The EUI is the measure of the total kWh used per year to the total number of square meters of conditioned space.	Site-specific reduction (%) compared to previous year EUI (Downward Trend)

Energy cost index (ECI): Measure of the total energy cost of a building over a period of time (year), per gross square meter of conditioned space.

Table 13.4 Energy Cost Index (ECI)

Value Driver	Unit	Measurement	Energy Reduction Target
Energy Cost Index (ECI)	$/m²/Year	The ECI adds up all costs of energy and divides the result by total square meters of conditioned space.	Site-specific reduction (%) compared to previous year ECI (Downward Trend)

Energy productivity index (EPI): Measure of the total energy consumed in a building over a period (year), per occupant or visitor, expressed in kWh/capita/year.

Table 13.5 Energy Productivity Index (EPI)

Value Driver	Unit	Measurement	Energy Reduction Target
Energy Productivity Index (EPI)	kWh/Capita/ year or $/Capita/ year	The EPI adds up all energy consumed (or billed) by the facility divided by the number of visitors or occupants.	Site-specific reduction (%) compared to previous year EPI (Downward Trend)

ENERGY CENTERED MAINTENANCE AND ENERGY PERFORMANCE INDICATORS

Energy performance indicators in energy centered maintenance model can be measured as part of the building's EUI, ECI and EPI. However, those indicators provide information about how the entire facility is performing from an energy perspective and not how each energy critical equipment is performing.

Therefore, new energy performance indicators need to be developed for each equipment by itself to measure its performance before and after the energy maintenance tasks took place and implemented. Those indicators differ according to equipment type and the type of energy it consumes (for example, measuring motor power consumption in an air handling unit before and after conducting the maintenance tasks, and measuring the effectiveness of water to water heat exchanger).

Building's energy performance indicators focus on energy consumption of the building, information about total energy consumption are extracted from energy bills or energy meters. On the other side, energy performance of equipment should be measured by the maintenance personnel while conducting energy centered maintenance tasks. Information about heat exchangers effectiveness need to be measured and calculated manually, and information about air handling unit's energy consumption can be extracted from a meter that is measuring that AHU motor energy consumption.

Energy performance indicators related to equipment performance involves collecting and analyzing actual data about the operational parameters of the equipment such as equipment's efficiency, equipment's effectiveness, power consumption, etc.

The following equipment's performance indicators can be used to measure the equipment's performance before and after energy centered maintenance tasks are conducted to measure the actual energy savings achieved:

Table 13.6 Performance Indicator for Mechanical/Electrical Equipment

Table 13.6.1 Air Handling Units:

Measurement	Performance Indicator	Target
Supply/ Return Airflow rate (m³/hr)	% of Actual Flowrate/ T&C Flowrate	± 5% of Testing and Commissioning Value
Full load motor's electricity consumption (kWh)	% of Actual Energy Consumption/ T&C Energy Consumption	± 5% of Testing and Commissioning Value
Full Load cooling coil pressure drop (PSI)	% of Actual Pressure Drop/ T&C Pressure Drop	± 5% of Testing and Commissioning Value
Full load chilled water temperature difference (°C)	% of Actual Delta T/ Design Delta T	± 2% of Design and Commissioned temperature difference (°C)

Table 13.6.2 Fan Coil Units:

Measurement	Performance Indicator	Target
Supply/ Return Airflow rate (m³/hr)	% of Actual Flowrate/ T&C Flowrate	± 5% of Testing and Commissioning Value
Full load motor's electricity consumption (kWh)	% of Actual Energy Consumption/ T&C Energy Consumption	± 5% of Testing and Commissioning Value
Full Load cooling coil pressure drop (PSI)	% of Actual Pressure Drop/ T&C Pressure Drop	± 5% of Testing and Commissioning Value
Full load chilled water temperature difference (°C)	% of Actual Delta T/ Design Delta T	± 2% of Design and Commissioned temperature difference (°C)

Table 13.6.3 Energy Recovery Units:

Measurement	Performance Indicator	Target
Supply/ Return Airflow rate (m³/hr)	% of Actual Flowrate/ T&C Flowrate	± 5% of Testing and Commissioning Value
Full load motor's electricity consumption (kWh)	% of Actual Energy Consumption/ T&C Energy Consumption	± 5% of Testing and Commissioning Value
Full Load cooling coil pressure drop (PSI)	% of Actual Pressure Drop/ T&C Pressure Drop	± 5% of Testing and Commissioning Value

Table 13.6.4 Boilers:

Measurement	Performance Indicator	Target
Fuel combustion	% of Actual Fuel Consumption Rate / T&C Fuel Consumption Rate	± 5% of Testing and Commissioning Value

Table 13.6.5 Pumps:

Measurement	Performance Indicator	Target
Full load water flow rate (gpm, lps)	% of Actual Flowrate/ T&C Flowrate	100% -110% of Testing and Commissioning Value
Full load motor's electricity consumption (kWh)	% of Actual Energy Consumption/ T&C Energy Consumption	± 5% of Testing and Commissioning Value

Table 13.6.6 Close Control Units:

Measurement	Performance Indicator	Target
Supply/ Return Airflow rate (m³/hr)	% of Actual Flowrate/ T&C Flowrate	± 5% of Testing and Commissioning Value
Full Load cooling coil pressure drop (PSI)	% of Actual Pressure Drop/ T&C Pressure Drop	± 5% of Testing and Commissioning Value
Full load Fan motor's electricity consumption (kWh)	% of Actual Energy Consumption/ T&C Energy Consumption	± 5% of Testing and Commissioning Value
Full load compressor's electricity consumption (kWh)	% of Actual Energy Consumption/ T&C Energy Consumption	± 5% of Testing and Commissioning Value
Full load chilled water temperature difference (°C)	% of Actual Delta T/ Design Delta T	± 2% of Design and Commissioned temperature difference (°C)
DX Unit Energy Efficiency Ratio	EER = Cooling Capacity (Btu per hour)/ Power Input (Watts)	Matching original EER as per Manufacturer records.

Table 13.6.7 Fans:

Measurement	Performance Indicator	Target
Supply/ Return Airflow rate (m³/hr)	% of Actual Flowrate/ T&C Flowrate	± 5% of Testing and Commissioning Value
Full load motor's electricity consumption (kWh)	% of Actual Energy Consumption/ T&C Energy Consumption	± 5% of Testing and Commissioning Value

Table 13.6.8 Cooling Towers:

Supply/ Return Airflow rate (m³/hr)	% of Actual Flowrate/ T&C Flowrate	± 5% of Testing and Commissioning Value
Full load motor's electricity consumption (kWh)	% of Actual Energy Consumption/ T&C Energy Consumption	± 5% of Testing and Commissioning Value
Cooling Tower Range at Full Load (°C)	In-Out Water Temperature	± 5% of Testing and Commissioning Value

Table 13.6.9 Air Cooled Chillers:

Measurement	Performance Indicator	Target
Full load evaporator pressure drop (PSI)	% of Actual Pressure Drop/ T&C Pressure Drop	± 5% of Testing and Commissioning Value
Full load condenser fans motor's electricity consumption (kWh)	% of Actual Energy Consumption/ T&C Energy Consumption	± 5% of Testing and Commissioning Value
Full load compressor's electricity consumption (kWh)	% of Actual Energy Consumption/ T&C Energy Consumption	± 5% of Testing and Commissioning Value
Full load chilled water temperature difference (°C)	% of Actual Delta T/ Design Delta T	± 2% of Design and Commissioned temperature difference (°C)
Energy Efficiency Ratio	EER = Cooling Capacity (Btu per hour)/ Power Input (Watts)	Matching original EER as per Manufacturer records.

Table 13.6.10 Plate Heat Exchangers:

Measurement	Performance Indicator	Target
Full load water pressure drop (PSI)	% of Actual Pressure Drop/ T&C Pressure Drop	± 5% of Testing and Commissioning Value
Full load water temperature difference (T hot-in – T Cold-in) (°C)	Difference between the inlet temperature on the hot water side minus the inlet cold water temperature on the cold side	± 2% of Design and Commissioned temperature difference (°C)

Table 13.6.11 Water Cooled Chillers:

Measurement	Performance Indicator	Target
Full load evaporator pressure drop (PSI)	% of Actual Pressure Drop/ T&C Pressure Drop	± 5% of Testing and Commissioning Value
Full load condenser pressure drop (PSI)	% of Actual Pressure Drop/ T&C Pressure Drop	± 5% of Testing and Commissioning Value
Full load compressor's electricity consumption (kWh)	% of Actual Energy Consumption/ T&C Energy Consumption	± 5% of Testing and Commissioning Value
Full load chilled water temperature difference (°C)	% of Actual Delta T/ Design Delta T	± 2% of Design and Commissioned temperature difference (°C)
Energy Efficiency Ratio	EER = Cooling Capacity (Btu per hour)/ Power Input (Watts)	Matching original EER as per Manufacturer records.

Table 13.6.12 Water Cooled Chillers:

Measurement	Performance Indicator	Target
Full load compressor's electricity consumption (kWh)	% of Actual Energy Consumption/ T&C Energy Consumption	± 5% of Testing and Commissioning Value
Full load chilled water temperature difference (°C)	% of Actual Delta T/ Design Delta T	± 2% of Design and Commissioned temperature difference (°C)
Energy Efficiency Ratio	EER = Cooling Capacity (Btu per hour)/ Power Input (Watts)	Matching original EER as per Manufacturer records.
Full load condenser fans motor's electricity consumption (kWh)	% of Actual Energy Consumption/ T&C Energy Consumption	± 5% of Testing and Commissioning Value

Table 13.6.13 Pressure Reducing Valve Stations:

Measurement	Performance Indicator	Target
Water pressure after PRV	% of Actual water pressure/ T&C water pressure record	± 5% of Testing and Commissioning Value

Table 13.6.14 Travelators, Lifts, and Escalators:

Measurement	Performance Indicator	Target
Full load motor's electricity consumption (kWh)	% of Actual Energy Consumption/ T&C Energy Consumption	± 5% of Testing and Commissioning Value

Table 13.6.15 Motor Control Centers:

Measurement	Performance Indicator	Target
In-out voltage records	In-out voltage readings	100% matching Testing and Commissioning Value in case no changes happened on electrical load connected to MCC

Table 13.6.16 Variable Frequency Drives:

Measurement	Performance Indicator	Target
In/out current at constant frequency	In/out current at constant frequency readings	100% matching Testing and Commissioning Value in case no changes happened on electrical load connected to MCC

SAVINGS IN DATA CENTER MEASURES AND VERIFICATION

If possible, a separate electricity meter for the data center will help measure the consumption and whether it is reduced or not through implementing energy reduction countermeasures. The main measure to identifying whether energy reductions occurred is the kWh per month consumption and the Electricity Intensity (kWh/Sq Ft. of the data center. The PUE (power usage effectiveness) is still a good measure of effectiveness, but these first two are the best for energy improvements verification.

Action 1. Place the servers in a hot/cold configuration with only cold air on one side and hot air coming from the server back on the other side.

Measures for Verification: kWh consumption per month and electricity intensity are the best verification measures.

Action 2. Check the temperatures in the server room. It is probably around 66 degrees Fahrenheit. The ASHRAE Standard allows significantly higher temperatures now than in the past. By raising the temperatures, less air conditioning is needed, less work on the cooling towers, and you may be able to retire one or more CRACs since they may not be needed anymore. Multiple data centers are found to be too cold and dry. In 2008 ASHRAE issued a revision of its guidance around the temperature of supply air to 64.4°F being lowest acceptable and 80.6°F highest (up from its previous recommended high of 77°F). Many data centers have been setting temperatures much lower, as low as 55°F. By raising the temperature even one degree, data centers can achieve energy savings. For humidity, ASHRAE has also loosened its guidelines, increasing the high end to 60% relative humidity (up from 55% in 2004).

Measures for Verification: kWh Consumption per month and Electricity Intensity are the best verification measures.

Action 3. Separate the cold air from mixing with the hot air on the way back to the CRAC. You may need a contractor to plan and have this built and put in place. Panels are used including panel doors to seal off the cold aisles. This project will have less than two years payback. Be sure to include the fire marshal and the maintenance department or contractor in on the beginning of the planning and execution.

Measures for Verification: kWh consumption per month and electricity intensity are the best verification measures.

Action 4. Using a wattmeter, find out for each server the watts being used to power the server and compare to the nameplate what should be used. The old servers could easily be "energy hogs" and if so, should be replaced.

Measures for Verification: Number of servers using excessive energy/total number servers inspected X 100 = percent servers inspected that are using excessive energy.

Action 5. Measure the utilization of each server. If the utilization is low, then have the programmers increase the use. This task will provide more server efficiency, thus reducing some servers for being used and lower the energy consumed. There will need to be some servers with lower utilization to be used as a backup to the main ones.

Measures for Verification: Servers Utilization = Sum of N1+N2+.... Nt/Total Number of Servers x 100

Action 6. Calculate either with actual data or estimated data, the P.U.E. (power usage effectiveness). Power usage effectiveness = total facility power (kW)/IT equipment power (kW). A PUE of 1.6 to 1.3 is good, with 1 being the goal. The lower the number, the larger percentage of the total power furnished to the facility is used by the data center operations. (80% of the power furnished used by the IT equipment would be very good. This is the DCIE indicator whose reciprocal is the PUE. 1/.8 = 1.25)

Measure for Verification: Power Usage Effectiveness(PUE) = Total Building Power (kW)/IT Equipment Power (kW) and then Goal-PUE = Y X100 = Percent Reduction Needed to Achieve the Goal

DEVELOPING AN ELECTRICITY BASELINE AND REDUCING ENERGY CONSUMPTION AND COSTS—A CASE STUDY

The aim of this part to illustrate to the reader how to conduct energy baseline for their facility and how to choose the equipment that must be part of energy centered maintenance model based on its energy classification process discussed in earlier sections.

The case study has been conducted on a multi-story commercial building, based on its electrical energy consumption. The facility has the following characteristics:

- Total build up area is 240,000.0 m²
- Total air conditioned area is 220,000.0 m²
- Building use: commercial building
- Building age: 4 years
- Cooling Source: District cooling (no chillers or cooling towers exist in the facility)

The common areas of the facility are served by multiple mechanical types of equipment as listed in the following table. The design connected load for each equipment in (kW) number of hours of operation each day (hr per day) are mentioned.

Table 13.7 Connected Electrical Load and Operating Hours

List of equipment's	Equipment Code	Qty	Design Power, kW	Operating hours, hrs/day
Primary chilled water pump	PCWP	4	95.0	16
Secondary Chilled water pump	SCWP	12	34.0	16
Car park supply fan	CPSF	5	30.0	16
Car park extract fan	CPEF	5	30.0	16
Fan coil unit	FCU	363	2.0	16
Fresh Air Handling Units 1	FAHU1	4	115	18
Toilet exhaust fan	TEF	4	6.0	18
Focus light	FL	24	1.0	6
Pole focus light	PFL1	64	0.15	6
Pole focus light	PFL1	128	0.07	6
Elevator motor	EM	31	17.5	10
Escalator motor	ESM	2	15.0	12
Water Booster Pumps	WBP	6	7.5	4
Water transfer pumps	WTP	4	30	6
Water feature pumps	WFP	10	7.0	6
Irrigation pump	IP	2	10.0	4
Fresh Air Handling Units 2	FAHU2	4	18.5	24
Fresh Air Handling Units 3	FAHU3	5	31.0	24
Air Handling Units 2	AHU	4	10.0	18

Energy Baseline

It is required to gain the electrical energy consumption of the facility for the past three years for a more accurate baseline. The data shall be extracted from real-metering or utility energy bills.

Based on three years' historical data, the recorded electrical energy consumption is shown in Table 13.8.

A bar chart will show the trends between the years' kWhs by month. The chart will tell you if the years are similar and if they are seasonal. Figure 13.3 shows that 2015 is lower consumption and 2013 is the highest consumption. All four seasons—spring, summer, fall and winter are evident in the graphs data.

Table 13.8 Energy Baseline

	Electricity consumption, kWh			Average Monthly Ambient temperature, Deg C		
Year	2013	2014	2015	2013	2014	2015
Jan	800,076	744,194	714,240	21	19	21
Feb	824,274	778,312	746,271	22	20	24
Mar	840,575	798,681	742,376	25	24	25
Apr	861,632	809,592	792,806	29	30	28
May	953,582	886,523	868,506	31	33	34
Jun	954,811	888,682	816,048	33	35	35
Jul	1,027,979	955,307	935,282	37	37	38
Aug	933,077	875,176	813,035	37	37	38
Sep	920,051	820,823	746,207	34	35	35
Oct	863,218	799,391	743,678	31	32	32
Nov	868,583	795,642	739,416	26	26	27
Dec	793,503	737,728	644,569	22	22	22
Total	10,643,374	9,892,065	9,304,449			

All years are identical in trend, but the consumption is not the same. The year 2013 shows the highest consumption compared to other years, while the year 2015 shows the lowest consumption.

In late 2013 some energy reduction projects and energy centered maintenance model were implemented and resulted in reduction happening.

The year 2013 was considered as baseline for two years' energy reduction program, the percent improvement in 2014 and 2015 when compared to 2013 are described in Table 13.9.

Energy Benchmarking

Energy benchmarking is ongoing review of building energy consumption to determine if the building energy performance is improving or not.

In our case study the benchmarking figures are developed by comparing 2013 energy utilization index (EUI) with the resulted values for

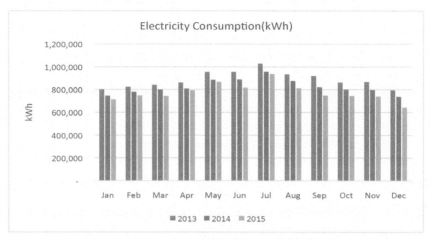

Figure 13.3 Electrical consumption trends by month

Table 13.9 Percent improvement compared to 2013

Total electricity consumption, kWh			Percent improvement compared to 2013		
2013	2014	2015	2013	2014	2015
10,643,374	9,892,065	9,304,449	baseline	(6.7%)	(12.2%)

years 2014 and 2015. This internal benchmarking compares the facilities energy performance with itself in previous years, but it doesn't compare it to other similar facilities. Therefore it is always recommended that the energy management team compares the facility energy performance with other similar buildings to decide if further improvements can be achieved.

As explained earlier, EUI is calculated based on total electricity consumed in the facility in a particular year, per total area of conditioned space.

The total air conditioned area of the facility is 220,000.0 m²

Table 13.10 Energy Utilization Index for Years 2013-2015

Total electricity consumption, kWh			Energy Utilization Index (kWh/m²/Year)		
2013	2014	2015	2013	2014	2015
10,643,374	9,892,065	9,304,449	48.37	44.96	42.29

Based on the EUI mentioned in above table, and by comparing it to other similar facilities, the energy management team shall be capable of deciding if the performance of the facility is acceptable from energy consumption perspective or not. Accordingly, the team should decide if another energy reduction program should be implemented or not, if so, a new energy reduction targets shall be set, and a new energy audit should take place to define potential areas of savings.

If external benchmarking is used as a reference, the energy management team should carefully collect all information and data about the facility you are comparing with. There are many factors can affect the energy consumption of the facility such as operating hours, type of air-conditioning system, the size of air conditioning units, the age of the facility, maintenance strategy and maintenance quality, control logic of building automation system, glass to wall ratio, etc.

The EUI in 2015 was 42.29 kWh/m²/year, and the energy utilization index for similar facilities is around 40.0 kWh/m²/year. Based on that, it can be concluded that the facility is operating in a close manner to other similar facilities. However, there is still potential areas of savings.

Energy Centered Maintenance Implementation
Step-1: Equipment Identification

Energy centered maintenance contributes to significant energy savings by ensuring energy critical equipment are operating efficiently.

The first step in this case study is to decide which equipment should be part of the energy centered maintenance analysis, which should be decided based on the given Energy Classification Code (ECC) for each equipment. The equipment will be assigned with and ECC of 5 to 1 based on the matrix and scale shown in Table 13.11.

The list of mechanical equipment serving the facility has been identified along with the connected electrical load and the total daily operating hours. Accordingly, the energy management team and maintenance personnel should be able to calculate the daily operating load in kW.hr.

The total daily operating load of all types of equipment is 51392.36 kW.hr per day. And the operating load for each equipment is also mentioned in the same table above. This information should be substituted in the following formula to decide the energy classification code for each equipment type.

Single Equipment Load Percentage % = Qty x Operating hours/day x Connected Power kW Full Day Load of all types of equipment kWh/day

For example, for primary chilled water pumps the load percentage should be:

Single Equipment Load Percentage % = 4 x 10 hrs/day x 95 kW 51392.36 kWh/day = 11.9%

Accordingly, and by referring to the energy classification scale, the ECC for it is 5.

Considering the same calculations the ECC for the rest of equipment is shown in Table 13.13.

The ECM model focus on all equipment with ECC of 5, 4 and 3 in its analysis, therefore, based on this case study, the following equipment will be selected for the analysis:
- Primary chilled water pump
- Secondary chilled water pump
- Car park supply fan
- Car park extract fan
- Fan coil unit
- Fresh air handling units 2
- Fresh air handling units 3

In this case study, we have selected fresh air handling unit 3 as an example to run the complete ECM model.

Step-2: Data Collection
The next step is to collect the operational parameters data about that equipment. Data can be collected from testing and commissioning figures, operation and maintenance manuals, asset registers, etc.

For fresh air handling units, the following data need to be gathered from testing and commissioning records:
- Supply/ return fan airflow rate (m³/hr).
- Motor power (Amps, Voltage).
- Cooling coil pressure drop (kPa).
- Cooling coil performance
- On coil/ off coil temperatures (°C).
- Chilled water delta T (°C).
- Air on/off heat wheel temperature (°C).

Table 13.11 Energy Classification Code

Energy Classification	Energy Impact	Description	Examples
High **4 or 5**	Large Impact Energy users	Systems with the following characteristics must be considered highly critical: • Operating Profile: Continuous Running. • Energy Capacity: High Capacity.	i.e. Chilled Water Pumps, AHUs Motors, Lighting, etc. - Efficiency loss on these systems will result in high associated energy costs. - Must be continuously running.
Medium **3**	Medium Impact Energy users	Systems with the following characteristics must be considered critical: • Operating Profile: Continuous Running. • Energy Capacity: Low Capacity.	i.e. Toilet Exhaust Fans, Lighting, etc. - Efficiency loss on these systems will have a medium impact on facility energy cost. - Might be running on time schedule
Low **1 & 2**	Low Impact Energy users	Systems with the following characteristics must be considered noncritical: • Operating Profile: Non-continuous Running. • Energy Capacity: Vary in Capacity.	i.e. Fire Pumps, Stair Case Pressurisation Fans, Emergency Lighting. - Might be high or low energy consumers - Operates only in case of emergency

Energy Classification Code

Figures 13.4 Energy Classification Code Scale

Table 13.12 Total Equipment's Operating Load

List of equipment's	Equipment Code	Qty	Design Power, kW	Operating hours, hrs/day	Operating load for each equipment kW.hr	Operating load for all equipment kW.hr
Primary chilled water pump	PCWP	4	95.0	16	950	3800
Secondary Chilled water pump	SCWP	12	34.0	16	340	4080
Car park supply air fan	CPSF	5	30.0	16	480	2400
Car park extract air fan	CPEF	5	30.0	16	480	2400
Fan coil unit	FCU	363	2.0	16	32	11616
Fresh Air Handling Units 1	FAHU1	4	115	18	2070	8280
Toilet exhaust fan	TEF	4	6.0	18	108	432
Focus light	FL	24	1.0	6	6	144
Pole focus light	PFL1	64	0.15	6	0.9	57.6
Pole focus light	PFL1	128	0.07	6	0.42	53.76
Elevator motor	EM	31	17.5	10	175	5425
Escalator motor	ESM	2	15.0	12	180	360
Water Booster Pumps	WBP	6	7.5	4	30	180
Water transfer pumps	WTP	4	30	6	180	720
Water feature pumps	WFP	10	7.0	6	42	420
Irrigation pump	IP	2	10.0	4	40	80
Fresh Air Handling Units 2	FAHU2	4	18.5	24	444	1776
Fresh Air Handling Units 3	FAHU3	5	31.0	24	744	3720
Air Handling Units 2	AHU	4	10.0	18	180	720
Total Daily Operating Load for all type of equipment (kW.hr/Day)						**51392.36**

Table 13.13 Calculated Energy Classification Code

List of equipment's	Equipment Code	Qty	Design Power, kW	Operating hours, hrs/day	Load Percentage %	ECC
Primary chilled water pump	PCWP	4	95.0	16	11.8%	5
Secondary Chilled water pump	SCWP	12	34.0	16	12.7%	5
Car park supply fan	CPSF	5	30.0	16	4.7%	3
Car park extract fan	CPEF	5	30.0	16	4.7%	3
Fan coil unit	FCU	363	2.0	16	22.6%	5
Fresh Air Handling Units 1	FAHU1	4	115	18	16.1%	5
Toilet exhaust fan	TEF	4	6.0	18	0.8%	1
Focus light	FL	24	1.0	6	0.3%	1
Pole focus light	PFL1	64	0.15	6	0.1%	1
Pole focus light	PFL1	128	0.07	6	0.1%	1
Elevator motor	EM	31	17.5	10	0.7%	1
Escalator motor	ESM	2	15.0	12	0.7%	1
Water Booster Pumps	WBP	6	7.5	4	0.4%	1
Water transfer pumps	WTP	4	30	6	1.4%	1
Water feature pumps	WFP	10	7.0	6	0.8%	1
Irrigation pump	IP	2	10.0	4	0.2%	1
Fresh Air Handling Units 2	FAHU2	4	18.5	24	3.5%	2
Fresh Air Handling Units 3	FAHU3	5	31.0	24	7.2%	4
Air Handling Units 2	AHU	4	10.0	18	1.4%	1

Those data define the baseline for the measurements in ECM model. The data collected are the expected operational parameters that the machine should deliver to meet its design intent.

For FAHU-3, the data are given in Table 13.14.

Step-3: Identify ECM Inspections, Frequency,
Craft, Tools, and Job Duration.

After collecting the required data, the maintenance personnel should plan for ECM inspection and decide what performance parameters need to be measured.

For FAHU-3, the following inspections and measurements shown in Table 13.15 should be conducted.

Step-4: Measuring Equipment Current Performance

The list of data that have been collected in step 2 are the recorded performance parameters of the equipment according to testing and commissioning records. The maintenance personnel will now measure the

Table 13.14 Technical Data for FAHU-3

Section	Element	Data
Heat Wheel Section	Supply air on heat wheel	28,000.0 m³/hr
	Return air on heat wheel	18,000.0 m³/hr
	air off-heat wheel temperature (dB/wB)	33.8 / 23.7 C
	Motor connected load (kW)	0.20 kW
Supply Fan Section	Supply airflow rate	28,000.0 m³/hr
	Motor connected power	18.5 kW
	Fan Efficiency	73.0 %
Cooling Coil Section	Air on cooling coil temperature (dB/wB)	33.8 / 23.7 °C
	Air off cooling coil temperature (dB/wB)	13.8 / 13.0 °C
	Supply/ Return chilled water temperature	4.0 / 12.0 °C
	Water pressure drop inside cooling coil	35.0 kPa
Return Fan Section	Return airflow rate	18,000.0 m³/hr
	Motor connected power	11.5 kW
	Fan Efficiency	70.0 %

current performance of the machine to compare the current readings to the first testing and commissioning readings.

Based on the measurements taken, the following data found to be inefficient:

- Supply airflow rate from supply fan
- Supply fan motor power consumption
- Difference between supply/ return chilled water temperatures

Step-5: Identifying Corrective Preventative
Actions and Cost Effectiveness

The maintenance personnel is now aware of which parts of the FAHU is not performing according to design intent. Accordingly, the team shall conduct root-cause analysis to investigate the reasons of those deficiencies.

Step-6: Updating Preventative Maintenance Plans

When the corrective actions are implemented, it is recommended now to include those measures in the reliability preventive maintenance job plans, but those ECM tasks can be executed either annually or every six months based on the maintenance team analysis and equipment requirement.

Table 13.15 Inspection and Measurements for FAHU-3

Inspection	Tool	Craft	Duration
Measure airflow rate (m³/hr)	Anemometer	Mechanical Technician	20-30 min
Check motor's full load current (Amps)	Multi-meter	Electrician	10-20 min
Measure Cooling Coil Performance on Full Load (Air Off Coil Temperature °C)	Thermometer	Mechanical Technician	10-15 min
Measure Cooling Coil Performance (Pressure Drop, PSI)	Manometer	Mechanical Technician	10-15 min
Measure 2-way/3-way control valves response to space temperature	DDC Simulator	Control Technician	20-30 min
Measure chilled water temperature difference (Delta-T) °C	Thermometer	Mechanical Technician	15-20 min

For example, regular PPM plans has the following tasks:

- Cleaning FAHU filters
- Tightening fan belt
- Cleaning cooling coil
- Lubricating shaft bearing

And as a result of the ECM analysis, the following tasks should be implemented:

- Measure fan flow rate and confirm all dampers functioning as required (annually)
- Measure motor power consumption and ensure it is optimum
- Measure actual chilled water delta T and ensure it is matching design values.

Table 13.16 Equipment Current Performance Results

Section	Element	Original reading (T&C)	Current reading (T&C)	Acceptable Performance measures	Is the current reading acceptable
Heat Wheel Section	Supply air on heat wheel	28,000.0 m³/hr	28,100.0 m³/hr	Current Value = ± 5% of Testing and Commissioning Value	Yes
	Return air on heat wheel	18,000.0 m³/hr	18,100.0 m³/hr	Current Value = ± 5% of Testing and Commissioning Value	Yes
	air off-heat wheel temperature (dB/wB)	33.8 / 23.7 C	34.1 / 23.9 C	Current Value = ± 5% of Testing and Commissioning Value	Yes
Supply Fan Section	Supply airflow rate	28,000.0 m³/hr	24,310.0 m³/hr	Current Value = ± 5% of Testing and Commissioning Value	No
	Hourly motor operating load (kW.hr)	18.5 kW.hr	19.8 kW	Match value on data plate	No
Cooling Coil Section	Air on cooling coil temperature (dB/wB)	33.8 / 23.7 °C	33.8 / 23.7 °C	Current Value = ± 5% of Testing and Commissioning Value	Yes
	Air off cooling coil temperature (dB/wB)	13.8 / 13.0 °C	13.1 / 12.6 °C	Current Value = ± 5% of Testing and Commissioning Value	Yes
	Supply/ Return chilled water temperature	4.0 / 12.0 °C	4.0 / 9.5 °C	100% ± 5% achieving desired Delta-T	No
	Water pressure drop inside cooling coil	35.0 kPa	34.5 kPa	Current Value = ± 5% of Testing and Commissioning Value	Yes
Return Fan Section	Return airflow rate	18,000.0 m³/hr	18,100.0 m³/hr	Current Value = ± 5% of Testing and Commissioning Value	Yes
	Hourly motor operating load	11.4 kW.hr	11.5 kW.hr	Match value on data plate	Yes
Section	Element	Original reading (T&C)	Current reading (T&C)	Acceptable Performance measures	Is the current reading acceptable
	(kW.hr)				

Table 13.17 Problem, Effect, Root-Cause and Corrective Action for FAHU-3

Problem	Effect	Root cause	Corrective Action	Is it cost effective
Supply airflow rate from supply fan, original flow rate is 28,000.0 m³/hr, but current flow rate is 24,310.0 m³/hr	Low cooling in served area	Increase in external static pressure due to closed motorized damper that supposed to be normally open at one of the branches.	Maintenance team checks all motorized dampers connected to this machine and damper open/close status was rectified.	Calculated the associated total cost and found it to be cost effective
Supply fan motor power consumption is expected to be 18.5 kW.hr, but the measured motor consumption is 19.8 kW.hr	High motor consumption and increased energy cost	Due to increase in external pressure, the motor consumed higher current to deliver run the fan	Motor power consumption will be measured on regular basis to ensure its not consuming higher energy	Calculated the associated total cost and found it to be cost effective
Design chilled water delta	Low delta T	Two-way valve	Actuators status	Minor cost
Problem	**Effect**	**Root cause**	**Corrective Action**	**Is it cost effective**
T is 8.0 °C while measured delta T is 5.5 °C.	penalties applied	actuator found stuck on opened position.	corrected and currently working on auto mode according to control logic	lower than applied penalties

Chapter 14

Building Energy Centered Behavior Leading to an Energy Centered Culture

KINDS OF ORGANIZATIONS' CULTURES

Many consultants and authors of energy and environmental books and articles outline the need for employees or students engagement in energy consumption reduction, water consumption reduction, equipment failure prevention, solid waste reduction, and environmental stewardship. With the engagement, positive desired behaviors should result, and an appropriate culture will prevail. The cultures sought are an energy awareness culture, a water awareness culture, a recycling culture to save the planet, an equipment failure prevention culture, and an environmental stewardship culture. With sustainability being implemented in the main businesses, universities, and colleges today, there is a need for all of these cultures be evident on site and practiced. These cultures together could be called a "sustainability culture" or an "energy centered culture."

CULTURE—DEFINITION AND BUILDING A SPECIFIC CULTURE

What is a culture? "It is the characteristics of a particular group of people, defined by everything from, religion, language, social behavior, manifestations, cuisine, music, and arts." http://www.livescience.com/21478-what-is-culture-definition-of-culture.html. Characteristics, knowledge, language, and group of people apply to the energy centered culture we desire to have in place to our business, college, or organization.

Figure14.1 Organizational Culture

Leadership—Management commitment is a must for success. Leaders are better when they have a vision or a policy.

Vision, Policy, and Goal—
    Vision—Reaching a state of betterment.
    Policy—A commitment of management what they are going to do or the employees are going to do.
    Goal—To improve something.

Values, Principles and A Standard—
    Management desires for all personnel to possess and follow on a daily basis.
    Examples: Committed to excellence, integrity, service before self, respect people, safety first.

Attitudes and Morale—What managers and employees believe and express to others. The morale is determined by whether the personnel is positive or negative or somewhere in between.

Communications—Uses of several different media to inform, make aware or teach organization's managers, employees, and contractors.

Roles, Responsibilities, and Measurement—Key players know what is respected of them and their authority. Have measures that show progress and results.

Business Processes—The way an organization gets its work done and achieves its mission and meets its customers' needs.

Training—**Awareness** and **Specialty Training**—Create an awareness of the culture that is needed and why, plus explaining how each can help and support energy reduction. develop skills in the organization's people that can help sustain energy management.

In other words, these eight attributes or characteristics are important to obtain the desired culture that remains in place as long as needed.

- Leadership
- Vision, Policy, Goals
- Values, Principles and Standard
- Attitudes and Morale
- Communications
- Roles, Responsibilities and Measurement
- Business Processes
- Training

All of the **characteristics** can help achieve the desired culture, but they are not of equal importance. Leadership, vision, policy, & goals, attitudes & morale, communications. Roles, responsibilities, management, and training, are essential to achieving the desired culture. These are interrelated such as good leadership can impact favorably everyone's attitude and morale. Vision or Policy can help motivated the **group of people** to join in and learn the **knowledge and language** and help accomplish actions that will contribute to goal attainment. Training is the method to depart the knowledge needed for every to perform specific desired tasks. Communications are the lifeblood that keeps the culture fluid, current on issues and tasks needed, plus knowing where current performance is and how it compares to the goals and vision. Roles, responsibilities, and management provide a structure, so everyone knows what their responsibilities are and what is expected of them.

Leadership with a vision or policy influences and changes attitudes, mindsets and behaviors of people and helps them share a purpose and achieve operations and process improvements.

The goal is to transition the workplace into everyone unconsciously showing positive behaviors regarding their energy and water use, environmental stewardship, observing and reporting leaks, noise, and other items that stop energy waste. The possible phases or stages of change is

Leadership Commitment
   Vision or Policy Developed—
   Good Positive Attitudes Evident—
   Roles and Responsibilities Defined—
   Both-way Communications—
Training and Commitment
   Implementing the Program's Objectives and Targets with Action
      Plans and Projects
   Measuring and Monitoring
   Taking Corrective Action (if needed)
   Recognizing & Rewarding Major Contributors to Program Success

Figure 14.2 Characteristics Needed for a Culture Change or Development

to go from the *state of being unconsciously and not participating* to become cognizant *of what to do* (through training) and being *consciously showing desired behavior* (driven by policies & procedures, being motivated, and being recognized and rewarded). Now the final stage of *being unconsciously but exhibiting positive and desired behavior automatically* comes from the training to know what to do, their commitment, the practice of doing it over and over and knowing they will be recognized and possibly rewarded for their participation.

The "What Behavior is Desired" comes through training. The main what in each of the cultured areas will be covered now. This list could be massive. Only four of the most common will be shown to get the point across how important it is that the participants know what is expected and what to do.

Water Consumption Reduction
1.   Report leaky faucets to maintenance or facilities.
     • Five drips a second is a steady stream
     • One gallon of water = 15,140 drips
     • A steady drip waste one gallon every 15 minutes
     • A steady trickle wastes one gallon every four minutes

2.   Reduce shower time to 5 minutes or less.
     • 150 gallons of water can be saved in a month by lowering the shower time by one to two minutes

3. Stop water from continuously running while brushing your teeth or shaving.
    *   Save up to 8 gallons of water a day by turning off the tap while you brush your teeth and shave.

4. Do not keep the water running while you are washing your auto.
    *   A garden hose can use 15 liters of water per minute. That's 225 liters of water in just 15 minutes!

Energy Consumption Reduction
1. Turn off the lights when you leave the room.
2. Use IT power management on your computer monitor and CPU. Saves approximately $75 each in energy cost every year.
3. Unplug appliances when they are not in use.

The appliances, electronics, and equipment use electricity when plugged in even though they are turned off.

Practice Energy Conservation
1. Implement an energy conservation policy
2. Develop an energy conservation program and implement it.

Solid Waste Reduction Culture
1. Recycling paper, aluminum cans, plastics, and cardboard. Recycling one ton of papers saves 17 trees, 4000 kWh of electricity, 7000 gallons of water, over three cubic yards of landfill, and a lot of green house gas emissions.
2. Recycle waste.

After computers, monitors, fax machines, printers and electronics reached the end of life, recycle to save precious metals and keep them from going to the water table.

Environmental Stewardship
1. Do not spill any chemicals. Could get in water table if not cleaned up.
2. Do not pour chemicals or put old medicine down the sink or toilet. Could do damage.
3. Participate as an ECO rep.

Help train others in environmental stewardship.
4.   Serve on green teams.
Help reuse, recycle and improve environmental stewardship.

Equipment Failure Prevention Culture
1.   Look for equipment running that may not be required to be running.
Motors often run 24/7 when maybe only need to for 8 hours.
2.   Listen for machinery making excessive noise.
Could need lubricating or a bearing replaced.
3.   Look for lights with a lot of dust or other debris on them.
Clean and restore the lumens to what they should be.
4.   Look for oil or water leaks.
Could be a severe environmental problem or loss of productivity due to an equipment failure if not caught and repaired.

Energy Centered Maintenance Culture
1.   Develop specific ECM model for your building.
2.   Implement ECM model as part of your maintenance strategy.
3.   Educate and train the maintenance personnel about ECM model.
4.   Engage energy management team with maintenance team to optimize the energy consumption of the equipment.

Chapter 15

Conclusion

DESIGNING AND IMPLEMENTING ECM

ECM provides the basis for identifying multiple low or no cost operation and maintenance practices that reduce energy consumption and improves the operational efficiency of the equipment. ECM works on the concept of returning the equipment to its original operational parameters (as originally commissioned) which result in improving its energy efficiency and reduce its energy use.

Energy centered maintenance model is a real maintenance program that focuses on energy-related equipment such as air handling units, electrical motors, pumps, etc. That equipment should be identified based on their energy consumption, and information should be collected from the testing and commissioning data (T&C) to compare the equipment's current behavior with its original commissioning data.

The proper execution of energy centered maintenance model should be delivered via planned job plans that are defined by the equipment's preventive and predictive maintenance needs. Equipment efficiency is significantly increased when tasks are accomplished by the standard proactive plans. Job plans should be designed for each piece of equipment based on relevant factors such as the maintenance tasks, frequency, tools, duration and craft type. The cycle of designing and implementing the job plans is the main step in an energy-centered maintenance process.

The ECM process's objective is to increase the energy efficiency of the equipment in a cost effective manner using proven maintenance assessments and identifying maintenance related tasks that measure and improve the current operational behavior of the equipment.

Unlike reliability maintenance, the energy centered maintenance model does not intend to enhance equipment reliability or to prevent failures; the aim is to create maintenance tasks that prevent energy waste during equipment operation and to ensure it is delivering the intended function

Development and implementation of ECM model also have the following objectives:

- To provide educational practice of how energy consumption is related to maintenance.

- Improving maintenance regime to focus on the operational condition of the equipment.

- To identify any change in the equipment performance compared to testing and commissioning data.

- To identify improvements which can be made to increase equipment's efficiency.

- Optimizing the energy consumption of the equipment.

- Increase energy efficiency of the equipment through a low operation and maintenance cost.

- Ensure that equipment delivers the expected operational parameters as per the design intent.

- Reduce energy consumption of the facility.

- Reducing greenhouse gas emissions, mainly CO_2 emission and reducing carbon footprint, caused by energy consumption.

Years ago, after giving a speech on maintenance at an industrial engineering conference, Marv Howell was asked, "Is there any connection with maintenance and saving energy?" I had not been asked this question before. My answer is yes. If we paint a room white, it will take fewer lumens to serve the room. If we change a filter in an air conditioner on time, the air conditioner will not have to work as hard, thus reducing energy consumption. That question in the last 15 years has often been asked. Organizations that provide facility and equipment maintenance would like to advertise or be able to tell their client they save energy while providing their maintenance. They were hesitant until ECM came along. ECM shows a direct relationship between reducing energy consumption and performing maintenance tasks. They knew that

⇨ Poor maintenance of energy-using systems, including significant energy users, is a major cause of energy waste in the governmental and the private sectors.

⇨ Energy losses from motors not turning off when they should, steam, water and air leaks, inoperable controls, and other losses from inadequate maintenance are large.

⇨ Uses energy consumption excess or energy waste as the primary criterion for determining specific maintenance or repair needs.

⇨ Lack of maintenance tasks in measuring the operational efficiency of the equipment such as motor power consumption and equipment effectiveness.

ECM uses energy consumption excess or energy waste as the primary criterion for determining specific maintenance or repair needs. energy centered maintenance (ECM) was originated in 2012 when Marv Howell kept finding motors running 24/7 when they were only required to run 7-8 hours daily. Also, he observed switches stuck on equipment, sensors not working, building automation systems with operators not trained, data centers using servers that were energy hogs and cold air mixing wrongly with the hot air on the way to the CRAC (computer room air conditioner). He discovered that a maintenance program needed to address equipment using excessive energy. In 2015, Fadi Al Shakhshir was working on developing a maintenance procedure that addresses what kind of maintenance tasks should be conducted during planned maintenance job plans that focus on the energy consumption of the equipment and the related operational parameters. As a final result, this book was written.

There are seven recognized maintenance types counting energy centered maintenance. They are:

1. Breakdown or reactive maintenance (before 1950, manufacturing revolution)
2. Preventative maintenance (1951)
3. Periodic maintenance (1951)
4. Predictive maintenance (around 1951)
5. Total productive maintenance (1951 origin, the 1980s in the USA)
6. Reliability centered maintenance (1960s origin, 1978 became known)
7. Energy centered maintenance (2012)

Each has advantages and disadvantages. The only one whose primary thrust is to reduce excessive energy use is energy centered maintenance. It is usually used in conjunction with preventative maintenance and predictive maintenance. This book outlines the steps to implement ECM and the equipment that should be included to maximize your energy savings.

The equipment to be included are:

I.   Mechanical Systems

    1. Heating, Ventilation, and Air Conditioning System
- Air Handling Units.
- Fan Coil Units.
- Energy Recovery Units.
- Boilers.
- Pumps.
- Close Control Units.
- Fans
- Cooling Towers.
- Air Cooled Chiller.
- Water Cooled Chillers.
- Heat Exchanger.
- Direct Expansion Air Conditioning Units.

    2. Water Supply System
- Pumps
- Heat Exchangers.
- PRV Stations.
- Boilers.

    3. Drainage System
- Sump Pumps (Sewage).
- Sewage Treatment Plant.

    4. Storm Water Management System.
- Rain Water Pumps.

    5. Building Transportation System
- Elevators.
- Travelators.
- Escalators.

II. Fire Fighting Systems
- Fire Pumps.

III. Electrical Systems
- Motor Control Centers.

- Variable Frequency Drive (VFD).

IV. Building Management System
- 2 Way - 3 Way Valve Functionality.
- Differential Pressure Switch – DPS.
- Differential Pressure Transmitter – DPS.
- Airflow Meters.
- Velocity Meters.
- On Coil Temperature & Humidity Sensors.
- Off Coil Temperature & Humidity Sensors.
- Space Temperature and Humidity Sensors.
- Thermostat Functionality.
- Control logic for all equipment

Each facility shall identify all systems and equipment applicable to their building, the technical information about the equipment can be found in asset registers, equipment schedules, O&M manuals, as-built drawings, etc. This information will be used to define the baseline performance of the energy related equipment and to design a maintenance checklist and plan as part of the preventive and predictive energy centered maintenance strategy.

The different equipment selected impacts four general areas, each important to any facility. They are comfort, productivity, health, and safety.

- Comfort—HVAC (heating, ventilation, & air conditioning), elevators & escalators, economizers, and lights & electrical system.
- Productivity—building transportation system, building management system, air compressors.
- Health—drainage systems, storm water management system, & water supply system.
- Safety—fire fighting system, ventilation

There are definite steps in developing and implementing ECM. The **first step** is to identify the equipment to be included in ECM. Energy criticality based on two factors, type of building systems and equipment classification codes. The classification code is based on the amount of energy consumed by the equipment and its operation.

This step gives us the equipment classification codes that identifies if the equipment is included or not in ECM.

The **second step** is data collection and equipment operational baseline. All equipment that will be included in ECM must go through a data collection process. Data collection is needed to define the baseline for the equipment's operational parameters. The required data focuses on the operational parameters of the equipment. The type of data required is different for each type of equipment. The necessary data are found in testing and commissioning records, O&M manuals, as-built drawings and other maintenance records. An example of the data needed for air handling units is the fan flow rate (m^3/hr.), motor power (volts, amps), VFD Data, and control valve control logic, etc.

Step 3 is to identify ECM inspections, frequency, craft, tools, and job duration. The frequency of conducting ECM inspections differ based on the type of equipment and kind of maintenance. The inspections need to be carried out by qualified staff that is capable of conducting the inspections, recording the data, and making sound judgments in relations to the equipment behavior.

Maintenance records should include:

- A coherent equipment repair history.
- A record of maintenance performed on equipment.
- Cost of maintenance.
- Cost of energy.
- Replacement information.
- Modification information.
- Spare parts replacement.
- Diagnostic monitoring data (if available in BMS).
- Condition assessment.
- Energy efficiency records.
- Retro-commissioning records.

Maintenance records can be used for activities such as energy efficiency analysis, energy centered maintenance inspections, preventive maintenance tasking, predictive maintenance tasking, frequency planning, and life cycle analysis.

Specifying energy centered maintenance inspections should be established based on clear targets, for example in an air handling unit the target is to ensure the AHU is capable of delivering the required airflow rate as intended in the design stage, this target sets what maintenance inspections should be conducted.

The inspections should be prepared considering the following:

- Determine which deficiencies may have an impact on customer satisfaction to correct.
- Determine which deficiencies are the most cost beneficial to correct.
- Determine which deficiencies are the most performance efficient to fix.
- Determine which performance parameters are critical for equipment operation to correct.

Scheduling energy centered maintenance inspections should be performed in such a way that energy centered maintenance tasks are conducted in the proper sequence, efficiently, and within prescribed time limits.

The frequency of energy centered maintenance inspections could vary based on different parameters such as:

- Equipment operating life.
- Physical condition.
- Failure interval and failure rate.

The more frequent the maintenance inspections take place, the higher the cost but, the greater chances of maintaining equipment efficiency. Conversely the less frequent the inspection, the less the cost but, the higher chances of increased energy use and increased energy waste intervals which result in high corrective maintenance cost. A balance between energy centered maintenance inspection, frequency, cost, and equipment efficiency should be assessed while defining the optimum ECM frequency.

The energy centered maintenance strategy calls to perform ECM inspections as part of equipment's regular preventive maintenance job plans; it will be either based on the following:

- Annual basis:
  Where ECM inspections will be part of the annual PPM plans.
- Semi-annual basis:
  Where ECM inspection will be part of the semi-annual PPM plans.

Energy centered maintenance inspections and job plans should be performed by a team of appropriately qualified and experienced personnel to achieve safe, and efficient maintenance operations.

Experienced maintenance personnel should meet the following criteria:

- Understand general facility systems and equipment layout.
- Comprehend the purpose and importance of the facility's systems and equipment.
- Understand the effect of ECM work on the facility's systems.
- Assimilate industrial safety, including hazards associated with specific systems and equipment.
- Understand job-specific work practices.
- Comprehend maintenance policies and procedures.
- Be familiar with the personal protective equipment.
- Be capable of evaluating the performance of the equipment.

Identifying tools, and specialized equipment that are required to execute an efficient energy centered maintenance inspection is an essential process that should be planned during the development of ECM job plans and inspections. A controlled supply of the proper type, quality, and quantity of tools and special equipment serves to avoid delays in maintenance work activities and increase worker efficiency. Defining the right tools is essential to allow the maintenance personnel to measure the current performance of the equipment which determines if any energy waste is noticed and to identify if the equipment is under performing, over performing or performing as intended.

The maintenance team should have a documented program for the control and calibration of test equipment and tools that ensure the availability of calibrated specialized equipment and tools.

Step 4 Measuring Equipment's Current
Performance and Compare to Baseline

The maintenance personnel should be capable of measuring the current performance of the equipment during energy centered maintenance inspection. Current performance defines the actual operational condition of the equipment which will be compared to the baseline performance as recorded in the testing and commissioning phase.

Measuring the equipment's current performance involves collecting and analyzing actual data about the operational parameters of the equipment. Data such as equipment's efficiency, power consumption, will help in determining if any part of the equipment is not delivering its intended function. Which in its turn results in identifying what kind of corrective actions should be done to improve and restore the operational efficiency of the equipment.

The measurements shall be compared with the baseline value, and the maintenance personnel shall be capable of judging if the equipment is over performing, underperforming or performing as intended. For example, measuring motor running current may be acceptable if it is operating within the acceptable range compared to testing and commissioning values, or an AHU is delivering an acceptable range of airflow compared to original data.

Step 5 Identifying Corrective/Preventive Action & Cost Effectiveness

When the root cause of a particular equipment performance deficiency is known, the right corrective action and repair can be implemented. Energy centered maintenance model focus on identifying any operational deficiencies or energy waste, therefore the corrective action may include specific repairs or replacements, therefore the cost effectiveness of the repair need to be determined prior conducting the corrective action.

Identifying corrective action will help in improving the overall balanced, proactive maintenance strategy in the following:

- Identifying what need to be done to restore equipment performance and what are the expected results.
- Identifying preventive actions to prevent the problem from being happening again.
- Identifying which maintenance process needs improvement.
- Identifying if new processes need to implement.
- Identifying new training for maintenance personnel.
- Identifying all costs associated with implementation and determining cost effectiveness.

Cost effectiveness should be calculated for each ECM tasks as well as corrective and preventive actions, calculating cost effectiveness should count for all maintenance activities costs associated with those actions, as well as potential energy reduction as defined in a certain period.

The maintenance costs include multiple elements such as:

- Time of maintenance
- Labor costs
- Material and consumable cost
- Equipment cost

- Calibration cost
- Spare parts cost

Energy saved could be calculated by different ways for each type of equipment by itself. For example, energy saved by enhancing motor efficiency equals the different in motor's kWh before and after enhancement, this difference is then can be converted to cost saving over a defined period (can be months, years, or life cycle of the equipment) and can be compared to maintenance cost.

Restoring equipment performance is focusing on the following outputs:

- Maximize equipment's operational efficiency.
- Restoring equipment's energy efficiency.
- Reducing energy waste by the equipment.
- Lowers operation cost by reducing energy consumption.
- Restoring the original performance of the equipment.
- Improves equipment quality.
- Ensuring equipment is delivering its intended performance.
- Understanding the effect of equipment age, operating environment on its performance.
- Obtaining data for continuous improvement and high operational effectiveness.

Corrective/preventive actions are required to restore the performance of a failing equipment. It is necessary to verify that the corrective action was effective, not only in eliminating the cause of the failure, but also in restoring the equipment's performance.

Step 6 Updating PM Plans on CMMS

Once the energy centered maintenance approach has been established, regular use of the maintenance program should be implemented regarding scheduled preventive maintenance as well as predictive maintenance. The new maintenance program should be updated on the balanced proactive maintenance strategy to establish a proactive system that reviews the energy efficiency data of the equipment which helps to identify system operational performance before a major deficiency occurs.

The new maintenance program should be updated in the maintenance management system (CMMS) to ensure all energy centered maintenance tasks that are defined during ECM inspection are now part of

the preventive maintenance plans and predictive maintenance practice.
    Updating PM plans requires the following:
* Identifying ECM tasks and frequency.
* Identifying and match the appropriate skill sets to the tasks.
* Identifying the appropriate materials to the tasks.
* Identifying the appropriate tools/special equipment to the tasks.
* Identifying all other resources needed to perform the job.
* Upload completed job plans to current CMMS

    The job plans provide all the details regarding safety, environmental, and regulatory issues, as well as the operations, required downtime, affected components/systems, materials, labor, and tools required to do the work. The procedural part of the plan contains a task or a logical sequence of tasks, while each task consists of a list of steps.
    In summary, the technical steps in developing and implementing ECM are:

Step 1. Identify the equipment to be included in ECM.
Step 2. Data collection and equipment operational baseline.
Step 3. Identify ECM inspections, frequency, craft, tools, and job duration.
Step 4. Measuring equipment's current performance and compare to baseline
Step 5. Identifying corrective/preventive action & cost effectiveness
Step 6. Updating PM plans on CMMS

    The eight characteristics of a successful energy reduction program are shown next. Energy centered maintenance should be included in characteristic #6 as an energy efficiency measure.

CHARACTERISTICS OF A SUCCESSFUL
ENERGY REDUCTION PROGRAM

    Eight characteristics keep showing up at organizations that have been successful at reducing energy consumption and energy costs. They are:

1. Top management leadership is committed and involved in the energy reduction effort and becomes the program's GLUE (good leaders using energy).

2. Energy reduction is made a corporate priority.
3. Corporate goals are established and communicated.
4. The energy champion or energy manager or both along with their cross functional energy team select challenging strategies that include development of an energy plan and objectives & targets with actions plans.
5. Key performance indicators (KPIs) and key results indicators (KRIs) are employed and kept current and visible to measure & drive progress and results.
6. Sufficient resources are provided to fund or ensure adequate countermeasures are implemented to achieve the corporate goals.
7. An energy efficiency culture is achieved.
8. Sufficient reviews are conducted to ensure continuous improvement, compliance to legal requirements, and adequate communications provided to keep all stakeholders informed, motivated, and engaged.

References

BY CHAPTER

Chapter 1 Energy Reduction
Effective Implementation of an ISO 50001 Energy Management System (EnMS), Marvin T. Howell, Quality Press,2014
Energy Centered Management-A Guide to Reducing Energy Consumption and Cost, Marvin T. Howell, Taylor and Francis Group, 2015
Implementing Energy Efficiency for Measurable Results Modules 1 & 2, Association of Energy Engineers, On Line Energy Seminar, instructor Marvin T. Howell, 2016
Energy Conservation for any Organization Modules 1 & 2, Association of Energy Engineers, On Line Energy Seminar, instructor Marvin T. Howell, 2016
The Energy Efficiency Strategy—The Energy Efficiency Opportunity in the UK, The Department of Energy and Climate Change, 2012.

Chapter 2 History of Maintenance and Different Maintenance Types
Operations and Maintenance Guide: Release 3.0, Energy.gov Office of Energy Efficiency & Renewable Energy. https://www1.eere.energy.gov/femp/pdfs/OM_5.pdf
Energy Centered Maintenance Modules 1 & 2, Association of Energy Engineers, On Line Energy Seminar, instructor Marvin T. Howell, 2016

Chapter 3 The ECM Model
Energy Centered Maintenance, Association of Energy Engineers, On Line Energy Seminar, instructor Marvin T. Howell, 2016

Chapter 4 The ECM Process-Equipment Identification
Energy Centered Maintenance, Association of Energy Engineers, On Line Energy Seminar, instructor Marvin T. Howell, 2016
HVCA, Standard Maintenance Specification for Mechanical Services in Buildings, SFG 20.

Chapter 5 The ECM Process-Data Collection

PECI. 1999. Operations and Maintenance Assessments. Portland Energy
Conservation, Inc. Published by U.S. Environmental Protection
Agency and U.S. Department of Energy, Washington, D.C.

HVCA, Standard Maintenance Specification for Mechanical Services in
Buildings, SFG 20.

CIBSE Guide F—Energy efficiency in buildings 2012—A Guide to
Energy Audit—U.S. Department of Energy 2011; BS EN ISO
50001:2011Energy Management Systems.

Emaar Maintenance Functionality Map, 2016.

http://fmlink.com/articles/your-om-technical-library-facilities-check-
list (about source of data)

Chapter 6 The ECM Process-Inspections

Energy Centered Maintenance, Modules 1 & 2, Association of Energy
Engineers, On Line Energy Seminar, instructor Marvin T. Howell,
2016

CIBSE Guide F—Energy efficiency in buildings 2012

A Guide to Energy Audit—U.S. Department of Energy 2011

Chapter 7 The ECM Process-Measuring
Equipment Current Performance

Energy Indicators that Drive Performance, Modules 1 & 2, Association
of Energy Engineers, On Line Energy Seminar, instructor Marvin T.
Howell, 2016

http://www.oee.com/calculating-oee.html

HVCA, Standard Maintenance Specification for Mechanical Services in
Buildings, SFG 20

Energy Centered Management System—A Guide to Reducing Energy
Consumption and Cost, Marvin T. Howell

Chapter 8 The ECM Process-Identifying
Corrective Preventative Actions and Cost Effectiveness

Low Hanging Fruit and Chasing Too Many Rabbits, Modules 1 &
2, Association of Energy Engineers, On Line Energy Seminar,
instructor Marvin T. Howell, 2016

Chapter 9 The ECM Process-Updating Preventative Plans

Energy Centered Maintenance Modules 1 & 2, Association of Energy

Engineers, On Line Energy Seminar, instructor Marvin T. Howell, 2016

Energy Centered Management System—A Guide to Reducing Energy Consumption and Cost, Marvin T. Howell

Chapter 10 Relation Between LTD and Maintenance
ECM Relation Between LTD and Maintenance

HVAC Chilled Water Distribution Schemes, Continuing Education and Development, Inc. A. Bhatia.

Degrading Chilled Water Plant Delta-T: Causes and Mitigations, by Steven T. Taylor, 2002.

Chapter 11 ECM in Data Centers

Energy Centered Maintenance Modules 1 & 2, Association of Energy Engineers, On Line Energy Seminar, instructor Marvin T. Howell, 2016

Low Hanging Fruit and Chasing Too Many Rabbits, Modules 1 & 2, Association of Energy Engineers, On Line Energy Seminar, instructor Marvin T. Howell, 2016

Chapter 12 Measures of Equipment and
Maintenance Efficiency and Effectiveness

http://www.dundas.com/blog-post/kpi-vs-kri-the-difference-and-the-importance

Energy Indicators that Drive Performance, Module 1, Association of Energy Engineers, On Line Energy Seminar, instructor Marvin T. Howell, 2016

Energy Centered Maintenance Modules 1 & 2, Association of Energy Engineers, On Line Energy Seminar, instructor Marvin T. Howell, 2016

Harvard Manage Mentor. (n.d.). Gathering Performance Data. Retrieved October 21, 2009 from http://ww3.harvardbusiness. org/corporate/demos/hmm10/performance_measurement/ set_targets.html

Phillips, L., Gray, R., Malinovsky, A., Rosowsky, M. (April 2009). The Assessment Report: Documenting Findings and Using Results to Drive Improvement. Texas A&M University Retrieved 10/12/09 from http://assessment.tamu.edu/wkshp_pres/AssessReport_ UsingResults.pdf

PMMI Project. (August 2005). Target Setting—A Practical Guide. Retrieved October 21, 2009 from http://www.idea.gov.uk/idk/core/page.do?pageId=845670

PMMI Project. (August 2005). Target Setting Checklist. Retrieved October 21, 2009 from http://www.idea.gov.uk/idk/core/page.do?pageId=845670

Energy Star, https://www.energystar.gov/, What is energy use intensity (EUI)

Chapter 13 Building Behavior Leading to an Energy Centered Culture

http://www.livescience.com/21478-what-is-culture-definition-of-culture.html

http://www.psychologicalselfhelp.org/Chapter11.pdf

https://studyacer.com/question/changing-behavior-case-study-analysis-410367

http://search.aol.com/aol/image?q=organizational+change+models+and+theories&v_t=webmail-searchbox

Energy Centered Maintenance Modules 1 & 2, Association of Energy Engineers, On Line Energy Seminar, instructor Marvin T. Howell, 2016

Implementing Energy Efficiency for Measurable Results Modules 1 & 2, Association of Energy Engineers, On Line Energy Seminar, instructor Marvin T. Howell, 2016

Chapter 14 Verification of Energy Savings and SEE Up

Energy Indicators that Drive Performance, Modules 1 & 2, Association of Energy Engineers, On Line Energy Seminar, instructor Marvin T. Howell, 2016

Energy Centered Maintenance Modules 1 & 2, Association of Energy Engineers, On Line Energy Seminar, instructor Marvin T. Howell, 2016

Chapter 15 Conclusion

Energy Centered Maintenance Modules 1 & 2, Association of Energy Engineers, On Line Energy Seminar, instructor Marvin T. Howell, 2016

ALPHABETICALLY

A Guide to Energy Audit—U.S. Department of Energy 2011

BS EN ISO 50001:2011Energy Management Systems.

CIBSE Guide F—Energy efficiency in buildings 2012

Degrading Chilled Water Plant Delta-T: Causes and Mitigations, by Steven T. Taylor, 2002.

Effective Implementation of an ISO 50001 Energy Management System (EnMS), Marvin T. Howell, Quality Press,2014

Energy Centered Management-A Guide to Reducing Energy Consumption and Cost, Marvin T. Howell, Taylor and Francis Group, 2015

Energy Centered Maintenance Modules 1 & 2, Association of Energy Engineers, On Line Energy Seminar, instructor Marvin T. Howell, 2016

Energy Indicators that Drive Performance, Module 1, Association of Energy Engineers, On Line Energy Seminar, instructor Marvin T. Howell, 2016

Emaar Maintenance Functionality Map, 2016.

Harvard Manage Mentor. (n.d.). Gathering Performance Data. Retrieved October 21, 2009 from http://ww3.harvardbusiness. org/corporate/demos/hmm10/performance_measurement/ set_targets.html

HVCA, Standard Maintenance Specification for Mechanical Services in Buildings, SFG 20.

HVAC Chilled Water Distribution Schemes, Continuing Education and Development, Inc. A. Bhatia.

http://www.oee.com/calculating-oee.html

http://search.aol.com/aol/ image?q=organizational+change+models+and+theories&v_ t=webmail-searchbox

https://studyacer.com/question/changing-behavior-case-study-analysis-410367

http://www.livescience.com/21478-what-is-culture-definition-of-culture.html

http://www.psychologicalselfhelp.org/Chapter11.pdf

HVCA, Standard Maintenance Specification for Mechanical Services in Buildings, SFG 20

Implementing Energy Efficiency for Measurable Results Modules 1
& 2, Association of Energy Engineers, On Line Energy Seminar,
instructor Marvin T. Howell, 2016

Low Hanging Fruit and Chasing Too Many Rabbits, Modules 1 &
2, Association of Energy Engineers, On Line Energy Seminar,
instructor Marvin T. Howell, 2016

Operations and Maintenance Guide: Release 3.0, Energy.gov Office of
Energy Efficiency & Renewable Energy https://www1.eere.energy.
gov/femp/pdfs/OM_5.pdf

PECI. 1999. Operations and Maintenance Assessments. Portland Energy
Conservation, Inc. Published by U.S. Environmental Protection
Agency and U.S. Department of Energy, Washington, D.C

Phillips, L., Gray, R., Malinovsky, A., Rosowsky, M. (April 2009). The
Assessment Report: Documenting Findings and Using Results to
Drive Improvement. Texas A&M University Retrieved 10/12/09
from http://assessment.tamu.edu/wkshp_pres/AssessReport_
UsingResults.pdf

PMMI Project. (August 2005). Target Setting—A Practical Guide.
Retrieved October 21, 2009 from http://www.idea.gov.uk/idk/
core/page.do?pageId=845670

PMMI Project. (August 2005). Target Setting Checklist. Retrieved
October 21, 2009 from http://www.idea.gov.uk/idk/core/page.
do?pageId=845670

Glossary

Annual Work Plan

The proactive maintenance schedule on a 52-week calendar. It lists every facility, system, and equipment with its job plan and frequency for each proactive maintenance activity to be performed.

Asset

Any facility, system, equipment, or component.

CMMS

Computerized Maintenance Management System

Corrective Action

Repairs made when an asset fails to operate as intended.

Craftsperson

Any qualified technician assigned to handle problem calls and PM procedures.

Equipment

The individual components of mechanical and electrical systems that are serving the building to function. Examples are heating, ventilating, and air conditioning (HVAC) systems; elevators; and communications systems.

Facility

The buildings, utilities, structures, and other land improvements associated with a building, operation or service.

Failure

Event rendering equipment non-useful for its intended or specified purpose during a designated time interval.

Flow meter

A device used to measure the flow rate of a liquid or air

Functional Failure

The loss of function as the inability of an asset to meet a desired standard performance.

Job Plan

It provides all the details regarding operations, materials, labor, and tools required to do the work.

Maintenance Management

The administration of a program using such concepts as an organization, plans, procedures, schedules, cost control, periodic evaluation and feedback for the effective performance and control of maintenance with adequate provisions for interface with related disciplines such as health, safety, environmental compliance, quality control, and security.

Maintenance Supervisor

The individual having authority and responsibility for specific maintenance activities at a facility or system.

Predictive Maintenance

Condition-based maintenance strategy where one or more equipment outputs is measured about the degradation of a component or subsystem.

Preventive Maintenance

Time, usage or cycle-based maintenance strategy in which periodic testing, servicing, adjustments, lubrication or inspections are performed on equipment to determine the progress of wear in components or subsystems. It can prevent/mitigate failure or detect hidden failures.

Planning

The identification and assessment of needed resources and the order in which the resources are required to complete a job plan in the most efficient manner. Planning defines the scope of work on a work order, the resources needed to complete the job, the sequence of the jobs, how long each job will take, and which jobs can be done concurrently.

Potential Failure

Point at which the facility, system, or equipment has been detected as failing.

Productive Work

Work that corresponds to a work order or is related to a piece of equipment.

Reactive Maintenance

Maintenance strategy where equipment is allowed to operate with no associated maintenance program. Also referred to as "Run-to-Fail."

Root-Cause Analysis (RCA)

The methodology used to identify solutions to prevent failure from occurring. It is not root causes that are sought: it is effective, controllable, goal meeting solutions to prevent failures.

RCA process is also used in RCM model to prevent operational deficiencies from occurring and to eliminate or reduce energy waste.

Reliability-Centered Maintenance (RCM)

A structured/logic-based process used to develop complete system and equipment maintenance programs providing the highest level of equipment reliability at best possible cost.

Scheduling

The assignment of job plans to a specific period of time to maximize the use of available resources.

Scheduled Work

Work that can be identified, predicted, or planned well in advance.

Scheduling

The assignment of definite amounts of work to personnel based on estimates of how many personnel labor hours are available for the planning horizon.

System

A logical and systematic group of assets that are necessary to support the facility's mission. A system must be described in a breakdown structure for each site so that it be properly identified and managed.

Site

All structures and systems that support a building or a facility within a facility, e.g., utilities, parking lots, roadways, bridges, fences, tunnel, etc.

Tools

Inventoried implements used to perform or assist in performing maintenance work functions within the facility, e.g. specialized hand tools, calibration tools, power tools, electric cords, mechanical tools, etc.

Tasks

Instructions to be followed in the performance of maintenance procedures.

Wrench Time

Productive work. This work is the actual mechanical work performed by a technician, manager or contractor.

Index